Kornberg's Operational Guideline Series in
Occupational Medicine

**VOLUME 1**

# The Workplace Walk-Through

**James P. Kornberg, M.D., Sc. D.**

**With foreword by**

**O. Bruce Dickerson, M.D., M.P.H.**

CRC Press
Taylor & Francis Group
Boca Raton  London  New York

CRC Press is an imprint of the
Taylor & Francis Group, an **informa** business

CRC Press
Taylor & Francis Group
6000 Broken Sound Parkway NW, Suite 300
Boca Raton, FL 33487-2742

© 1992 by James P. Kornberg
CRC Press is an imprint of Taylor & Francis Group, an Informa business

First issued in paperback 2019

No claim to original U.S. Government works

ISBN-13: 978-0-367-45035-9 (pbk)
ISBN-13: 978-0-87371-620-8 (hbk)

**Visit the Taylor & Francis Web site at**
**http://www.taylorandfrancis.com**

**and the CRC Press Web site at**
**http://www.crcpress.com**

## DEDICATION

This book is dedicated first and foremost to my wife, Sally, daughters Mariah, Jamie and Terra, and to my numerous teachers throughout the years, especially, Mr. Joe Hale.

A very special dedication is also gratefully extended to my mother, brother, Sandy, and to my father, whose spirit and wisdom have guided me throughout this entire endeavor.

**James P. Kornberg, M.D., Sc.D.** graduated from the Massachusetts Institute of Technology with a B.S. and M.S. in Aeronautical and Astronautical Engineering. He then went on to Harvard University to earn a Doctor of Science in Environmental Engineering. He later enrolled in Dartmouth Medical School where he earned his M.D. degree. Following internship in a Columbia University teaching hospital, he completed his residency training in Occupational Medicine at the Harvard University, School of Public Health. While in the Boston area, Dr. Kornberg also worked as a senior staff consultant in Occupational Medicine for Arthur D. Little, Inc. in Cambridge, Massachusetts.

Dr. Kornberg is board-certified in Occupational Medicine under the American Board of Preventive Medicine and is a fellow in the American College of Preventive Medicine. He served four years on the Board of Directors of the Rocky Mountain Academy of Occupational Medicine. He currently serves as the medical advisor to the Colorado Mining Association and has been recognized as an expert witness at both the federal and state court levels in occupational medicine, occupational lung disease, and clinical toxicology. His clinical and consulting practice in occupational medicine includes working as consulting medical director or advisor to numerous diverse, local and national corporations which are headquartered in the front range area of the Rocky Mountains.

In August of 1990 and again in January of 1991, during the Persian Gulf War, Dr. Kornberg acted as a medical consultant and seminar leader to various agencies and institutions in Israel, including the Israeli Army, Hebrew University, and the Israeli Department of Environmental Health, regarding chemical disaster preparedness and the treatment of chemical and biological warfare exposures.

At present, Dr. Kornberg is President and Medical Director of COHBI Corporation (Comprehensive Occupational Health for Business and Industry) located in Boulder, Colorado.

# SERIES PREFACE

## KORNBERG'S OPERATIONAL GUIDELINE SERIES IN OCCUPATIONAL MEDICINE

I am pleased to introduce a long awaited and long overdue series of practical guidelines in occupational medicine which is designed specifically for the physician, who must respond to the challenges of occupational medicine practice and may not have had the time or the experience to prepare for the task at hand.

Volume I is the first of a series of several volumes to come, each of which will provide the physician with more advanced tools for performing not only the routine tasks of occupational medicine but also the more unusual and challenging assignments. Guidelines in the series will cover critical areas of practice, which are poorly treated or not covered at all in *traditional textbooks* in the field (e.g., the workplace walk-through, occupational medical causality analysis, the design of medical surveillance programs and even pass/fail criteria for respirator certification).

Each volume will be designed to be concise, understandable, affordable, and usable. Generally speaking, most volumes will also contain specific forms and/or checklists which may be utilized or customized by the reader.

The overall purpose of this new series is to improve the practice of occupational medicine. This goal is achieved by cultivating and educating the specialist and non-specialist alike, regarding the contemporary challenges of occupational medicine, while simultaneously equipping each practitioner with a better set of tools with which to perform both as a clinician and as a consultant on a daily basis. This series will usually be practical in design; and although it occasionally may be theoretical, it will always be relevant. The reader, moreover, may be confident with the knowledge that virtually all of the tools, constructs, and ideas provided in this series have been successfully implemented in the specialty practice of occupational medicine over the past 12 years.

James P. Kornberg, M.D., Sc. D.
Boulder, Colorado
February 1992

## PREFACE TO VOLUME I

## THE WORKPLACE WALK-THROUGH
## -OPERATIONAL GUIDELINES FOR THE PHYSICIAN-

Though important historically, the workplace walk-through is now and will be an even more essential activity for any physician who practices occupational medicine. Major changes are on the way, such as the Americans with Disabilities Act (ADA),[1] and the increasing emphasis upon industrial hygiene and safety instruction in some occupational medicine residency training programs. In addition, there are the expanding demands placed upon the occupational medicine specialist by attorneys, judges, insurance carriers, and employers to support theories of cause and effect, placement decisions, and medical restrictions with facts, not fantasies or guesses, about workplace exposure scenarios. Gone will be the days when the physician (specialist or otherwise) can utilize poorly conceived and undocumented, second or third-hand assertions about the workplace to buttress major decisions which will affect the worker, the employer, and other parties so profoundly. Under ADA, moreover, the physician can expect to play an even more important role in deciding worker fitness, and the nature and scope of reasonable accommodations for disabled workers, while simultaneously helping the employer to define the essential functions of a given job. Performance of these roles cannot be achieved over the telephone or by sitting in one's office. They must be achieved by visiting and becoming familiar with the details of the workplace as a total physical and functional environment in which the worker-patient spends a large percentage of time.

The capable performance of these expanded duties should propel occupational medicine onto an even higher plane of respect than it enjoys today. None of these laurels will accrue, however, unless all practitioners of occupational medicine take these challenges seriously.

Volume I of *Kornberg's Operational Guideline Series in Occupational Medicine* is dedicated to providing the physician or occupational health specialist with direction in four critical areas of practice. These are:

- How to perform a workplace walk-through.

- How to organize occupational health and industrial hygiene information (including that gathered during the walk-through) into an integrated and cohesive framework from which medical surveillance recommendations can be made (PATHMAX™; Appendix I).

- How to perform an occupational medical causality determination (Appendix II).

- How to begin to understand and to utilize selected environmental engineering and industrial hygiene concepts and methods, while providing direction for further learning with abundant references (Appendix VI).

Appendix III outlines a few special considerations which apply to performing the walk-through for a smaller company or employer.

A walk-through checklist summary is provided in Appendix IV to facilitate the collection of information and to avoid omissions of important topics.

Appendix V provides the physician with an inventory of possible, additional topics which can be discussed with the employer before the walk-through commences. These topics are highly specialized and may apply only during the focused investigation of specific problems.

Appendix VI is a technical companion to this volume which should prove to be a useful reference to those who are forever hoping to find a wide variety of critical definitions and a few important equations, well documented and all in one place. For those who have never appreciated the role of the industrial hygienist or safety engineer,

Appendix VI will overhaul their opinion quickly. The mastery of this appendix will be just the first step in learning the technical language that the engineering side of occupational safety and health depends upon everyday.

Finally, Volume I contains a handy glossary which should demystify some of the basic science of the walk-through and facilitate one's reading of Appendix VI.

Volume I was written as a humble, yet ambitious, effort to cover several critical and complicated subjects. While composing this volume, for example, it became evident that in conjunction with addressing the workplace walk-through, Volume I also appeared to be the ideal place to introduce both PATHMAX™ and a rational set of rules for conducting an occupational medical causality analysis. The addition of Appendix VI, moreover, was made to encourage the reader to use this volume as a somewhat customized technical reference. In summary, therefore, the overall structure, format and content of Volume I evolved with several different objectives in mind.

Ambitions aside, it became clear in finalizing Volume I that in order to maintain the *Operational Guidelines* format and purpose of the overall series, not all aspects of the topics in Volume I could be covered completely. The author accepts any shortfalls graciously and simply hopes that what has been provided is found to be both useful and motivating to the reader.

JPK

# ACKNOWLEDGMENTS

The impetus to embark upon the first volume of my new Operational Guideline Series in Occupational Medicine must be credited, in part, to my friends and colleagues, John R. Goldsmith, M.D., M.P.H. at Ben Gurion University in Beer Sheva, Israel, and Elihu D. Richter, M.D., M.P.H., Head, Unit of Occupational and Environmental Medicine at Hebrew University in Jerusalem, Israel. The suggestion by Drs. Goldsmith and Richter to *write an article for the general physician on how to evaluate an industrial setting* led directly to my focusing upon this task as the main topic of my first volume.

The portions of this volume related to occupational causality analysis and PATHMAX™ were developed by me within the fertile collegial milieu provided by my close medical associates, Michael R. Striplin, M.D. and Mark Bradley, M.D., M.P.H., along with my generous and patient legal tutor, Robert A. Weinberger, J.D. Manuscript preparation, critical review suggestions, moral support, and editorial excellence were provided by two distinguished members of my staff, Linda A. McKeever and Margaret E. Poyton, C.P.A.

Finally, I gratefully acknowledge the other members of my staff at COHBI Corporation (Comprehensive Occupational Health for Business and Industry), including my professional associate, Loretta G. Milburn, M.D., all of whom demonstrated exceptional patience and support while the preparation of this volume often took me away from my routine duties.

# FOREWORD

Over the last two decades significant and far-reaching changes have occurred in the fields of occupational medicine and environmental health. Not only have we seen the emergence of strong governmental regulatory agencies overseeing and monitoring our field; but matching this movement have been the rise of advocacy groups seeking stronger commitments for clean environments and worker safety, the embodiment of risk evaluation and risk acceptance into decision-making, and the use of the legal process to resolve differences. These changes have created an overwhelming requirement for physicians to assimilate and to organize large quantities of information on environmental and occupational hazards in order to maintain their traditional role as chief health consultant to the population on these matters.

In addition, in response to this challenge, we have seen an overlaying of our traditional training to include such subjects as occupational and environmental epidemiology, risk communication and management, systems safety, industrial hygiene, health physics, and ergonomics. Thus, the information overload continues.

However, as rapid as our professional advances have been, and as great as our efforts to meet these challenges have progressed, both new and established physicians, alike, must, nevertheless, continue to strive to make sense out of the swirling quagmire of information with which we are faced.

In this regard, it is heartening to see this highly commendable effort by James P. Kornberg, M.D., Sc. D., a leading occupational medicine specialist, to bring organizational methodology to help deal with the shower of environmental and occupational information that is falling upon us as physicians. With this first volume of his series, Dr. Kornberg places into the hands of physicians and risk managers, alike, a remarkable new tool, containing operational guidelines and technical recommendations, that allows the physician to perform a workplace walk-through, along with the necessary occupational and environmental health analyses of plants and work areas.

Unique to this first volume is PATHMAX™, a methodology incorporating informational and correlative matrices, which enable the physician to assemble and organize a large quantity of legislative, occupational hazard, job component, and health surveillance information into a logically sequenced format. This novel and long awaited information system provides the critical tools with which we, as physicians, can identify potential risks and can promulgate recommendations for risk management.

After reviewing a copy of Dr. Kornberg's first volume, I made immediate use of it in my practice; and I look forward to reviewing Volume II (Methods in Occupational Medicine), Volume III (Decision-Making in Occupational Medicine), and future volumes at the earliest opportunity.

O. Bruce Dickerson, M.D., M.P.H.

Medical Director, Southern New England Region,
Volunteer Hospitals of America
Past President, American College of Occupational Medicine
and American College of Preventive Medicine
Former Corporate Director of Health and Safety,
IBM Corporation

# TABLE OF CONTENTS

# LIST OF TABLES

# LIST OF FIGURES

# The
# Workplace
# Walk-Through

# INTRODUCTION

With increasing frequency over the past decade or so, both occupational medical specialists and non-specialists alike have been called upon by plant managers, industrial hygienists, safety personnel, and others to perform on-site visits to the workplace as part of their routine medical duties.[2-6] These visits are commonly called *workplace* or *plant walk-throughs*. Requests for such visits have developed for a variety of different reasons; but all of these reasons have, in common, the evolving role of the physician, both as a manager[7-10] and as a clinician, within the field of occupational medicine.

Despite some official precedents[11] stipulating that the physician must be familiar with the physical work environment, many physicians find themselves to be unprepared or insecure in assuming such responsibilities, because of inexperience, minimal training, or fear that they have little to offer. Some physicians, moreover, do not fully understand the manner in which their contributions and observations can be utilized in the development and management of the overall occupational health program.

The purposes of this volume are to:

- identify for both occupational medical specialists and non-specialists alike the majority of circumstances under which and the purposes for which walk-throughs are requested.

- provide some general guidelines regarding the character and extent of information requested by clients.

- illustrate an important methodological system (PATHMAX™; Appendix I) within which gathered data can be practically utilized.

- outline a logical construct (Appendix II) which defines the fundamental reasoning skills necessary to draw conclusions about occupational medical cause and effect relationships, based upon data gathered during the walk-through.

- warn the reader about the limitations of the walk-through and about those client requests which should be avoided.

The methods and details provided within this volume are not intended to apply and cannot be applied to all workplaces (especially smaller ones) which the physician may be asked to evaluate. Certain caveats to consider, therefore, when approaching the smaller workplace, in particular, are outlined in Appendix III.

The main goal in this volume is to provide targeted methodology for evaluating medium (250-500 workers) to large (>500 workers) facilities. It should be emphasized, however, that the general principles outlined within this volume definitely apply to smaller (perhaps, more commonly visited) workplaces: and the physician responsible for the walk-through, therefore, can confidently utilize such principles without reservation.

Finally, before proceeding, it should be emphasized and acknowledged that under many circumstances, the occupational health nurse is the leader of the occupational health team[6,12] and may be called upon to perform the workplace walk-through. The guidelines in this volume, therefore, are also designed for this professional and should be applied within the scope of the nurse's experience and training.

## THE WALK-THROUGH: OPERATIONAL GUIDELINES

### Indications for the Walk-Through

The reasons for which a walk-through may be requested can be arranged into the following categories:

■ **Preventive Services**

- Assessment of existing *program status* by adherence to established *walk-through guidelines* and *walk-through imperatives*

- Creation of a data base for *program development* or modification

■ **Investigative Services**

- Analysis of a potential workplace *causal relationship* between a suspected or documented workplace exposure and a suspected or diagnosed medical problem

- Data gathering as part of an overall *risk management* or *cost-containment* effort for the employer

■ **Regulatory or Audit Services**

■ **Formulation of Accommodation Options for Disabled Workers**

**Preventive Services**

*Program Status*

It is not uncommon for industrial hygiene, safety, and management personnel to embark independently from the physician to establish a program of occupational safety and health. Once completed, these professionals may wish the physician responsible for worker diagnosis and treatment to *see the workplace*, so that he or she can obtain a better understanding of the worker's job requirements, at least to a degree beyond the understanding offered by the job description (assuming that one exists) or the occupational history provided by the worker. In other words, the physician is asked to perform a *workplace walk-through*.

The format and methodologies for conducting the walk-through vary greatly, since there are no specific standards set for such an endeavor. A few guidelines, however, can be obtained from *institutional efforts* such as the Health Hazard Evaluation (HHE) conducted routinely by a NIOSH health and safety team, which often includes a physician.[13]

Given the absence of officially accepted walk-through guidelines, the following practical recommendations will be provided within the context of the task of evaluating *Program Status*. A physician's *Walk-Through Check List*, summarizing the practical recommendations discussed in this section, is provided to the reader in Appendix IV.

*Walk-Through Guidelines*

*The Pre-Walk-Through Meeting*:  Experience has shown conclusively that before beginning the actual physical walk-through of the workplace, the physician should request that an initial pre-walk-through orientation be conducted in an office or classroom setting. During this orientation:

- The purpose of the walk-through should be fully explained and the deliverable product to be provided by the physician should be identified and accepted by all parties.

- The overall operations of the work setting should be reviewed, with specific emphasis upon the location of key *functional operations* such as shipping/receiving, warehouse, production, quality assurance, clerical/management services, laboratory, and any other relevant categorical activities which may apply to the worksite under consideration.

It should be emphasized that, most often, large *functional operations* can be ultimately broken down into *unit operations* (those activities which are *static* in place and usually repetitive - e.g., soldering on an assembly line).  Some simple job activities consist of performing work within only one unit operation; but complex job activities, within large *functional operations*, always include performance of work within several different unit

## FIGURE 1

## THE RELATIONSHIP BETWEEN UNIT OPERATIONS AND JOB DESCRIPTIONS WITHIN AN OVERALL FUNCTIONAL OPERATION

### FUNCTIONAL OPERATION $F_1$

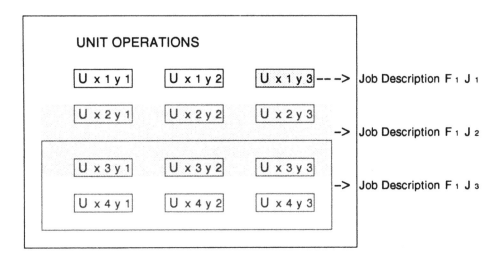

$F_1$ represents a single, overall *functional operation* such as shipping/receiving. $U_{X_iY_j}$ (i = 1-4; j = 1-3) represents a basic *unit operation* (e.g., labeling of packaged product for shipment) which is *static* or fixed in location. Job description $F_1J_1$ (shaded area) represents a simple job, involving work within only one unit operation ($U_{X_1Y_3}$). Job description $F_1J_2$ (larger shaded area) is also one job description (e.g., overall preparation of product for shipping) which encompasses six unit operations as shown. $F_1J_3$ is another job description (e.g., loading of product for shipping and unloading of raw materials upon receiving) which also encompasses six unit operations, three of which ($U_{X_3Y_1}$; $U_{X_3Y_2}$; and $U_{X_3Y_3}$) are in common with job description $F_1J_2$.

operations (e.g., performance of equipment maintenance, a job which is *dynamic* in place and usually not repetitive) (Figure 1). Initially familiarizing the physician with *functional and unit operations* is indispensable, since workplaces vary greatly in design and operation and trying to achieve the *big picture* while hurrying through the physical work environment is difficult, at best.

■ Job descriptions must be outlined and the relevance of each placed into the context of overall plant functional operations and unit operations.

■ Material Safety Data Sheets (MSDSs) should be presented and explained. The MSDS (Figure 2) is a fairly standardized document provided by a chemical user or manufacturer, which outlines a wide assortment of information about a specific chemical product or substance. The MSDS includes, among other information, the product's composition, handling requirements and anticipated health effects, along with brief guidelines for diagnosis and treatment following exposure. When reviewing MSDSs, it is important for the physician to make note of the trade names of products utilized by the company and to determine whether the information on the MSDS is specific enough to be useful.

Some manufacturers who formulate simple chemical components into proprietary mixtures will, for example, indicate on the MSDS only that the product contains *solvents*. This information may be inadequate for occupational safety and health purposes since, as in this example, the term *aromatic solvents* could refer to compounds which differ significantly from a toxicological point of view (e.g., benzene versus toluene).

■ Potential physical, biological, radiological, and ergonomic hazards must also be identified and discussed. Whenever possible, all potential hazards should be specified in relationship to identifiable unit operations. When unit operations cannot be specified, potential hazards should be correlated with job descriptions, functional operations, or even with specific physical areas within functional operations in the workplace.

- Health, safety and industrial hygiene policies and programs should be discussed in detail, sometimes along with certain *more delicate* issues relating to corporate priorities, management philosophy, and budgetary allocation for health and safety programs.

- Major current and historical health and safety problems along with the results of recent internal or external audits should be outlined, in conjunction with steps taken for remediation.

- Existing programs of data collection and recordkeeping should be discussed, along with standard procedures for data analysis and epidemiological forecasting.

- Additional topics may also be addressed such as those outlined in Appendix V, depending upon the workplace being toured.

*The Walk-Through*: Once the pre-walk-through meeting is completed, the tour begins. The physician should be certain that the proper safety equipment for the tour (e.g., eye or hearing protection, hard hat, steel-toed shoes, proper respirator, protective clothing, etc.) is either provided by the employer or brought, personally, to the meeting.

The physician should also bring a notebook in which to record all findings (aluminum encased is desirable). As a general rule, the physician should not engage any employee into extended conversation without first asking the employee's supervisor or the walk-through host. The reason for this recommendation is related to both safety considerations and etiquette. Once permission is obtained (and it almost always is), discussions with individual employees are encouraged, because they provide some of the most useful and insightful information available about virtually all *walk-through imperative* categories to be outlined (see page 10).

# 8 Operational Guidelines

Figure 2 (14)

| Material Safety Data Sheet | U.S.Department of Labor |
|---|---|
| May be used to comply with OSHA's Hazard Communication Standard, 29 CFR 1910.1200. Standard must be consulted for specific requirements. | Occupational Safety and Health Administration (Non–Mandatory Form) Form Approved OMB No. 1218–0072 |
| IDENTITY (As Used on Label and List) | Note: Blank spaces are not permitted. If any item is not applicable, or no information is available, the space must be marked to indicate that. |

## Section I

| Manufacturer's Name | Emergency Telephone Number |
|---|---|
| Address (Number, Street, City, State, and ZIP Code) | Telephone Number for Information |
| | Date Prepared |
| | Signature of Preparer (optional) |

## Section II – Hazardous Ingredients/Identity information

| Hazardous Components(Specific Chemical Identity: Common Name(s)) | OSHA PEL | ACGIH TLV | Other Limits Recommended | %(optional) |
|---|---|---|---|---|
| | | | | |
| | | | | |
| | | | | |
| | | | | |
| | | | | |
| | | | | |
| | | | | |
| | | | | |
| | | | | |
| | | | | |
| | | | | |

## Section III – Physical/Chemical Characteristics

| Boiling Point | | Specific Gravity (H2O = 1) | |
|---|---|---|---|
| Vapor Pressure (mm Hg) | | Melting Point | |
| | | Evaporation Rate | |
| Vapor Density (AIR = 1) | | (Butyl Acetate = 1) | |
| Solubility in Water | | | |
| Appearance and Odor | | | |

## Section IV – Fire and Explosion Hazard Data

| Flash Point (Method Used) | Flammable Limits | LEL | UEL |
|---|---|---|---|
| Extinguishing Media | | | |
| Special Fire Fighting Procedures | | | |
| Unusual Fire and Explosion Hazards | | | |

## Section V – Reactivity Data

| Stability | Unstable | | Conditions to Avoid |
|---|---|---|---|
| | Stable | | |

| Incompatibility (Materials to Avoid) |
|---|

| Hazardous Decomposition or Byproducts |
|---|

| Hazardous Polymerization | May Occur | | Conditions to Avoid |
|---|---|---|---|
| | Will Not Occur | | |

## Section VI – Health Hazard Data

| Routes(s) of Entry | Inhalation? | Skin? | Ingestion? |
|---|---|---|---|

| Health Hazards (Acute and Chronic) |
|---|

| Carcinogenicity: | NTP? | IARC Monographs? | OSHA Regulated? |
|---|---|---|---|

Signs and Symptoms of Exposure

Medical Conditions
Generally Aggravated by Exposure

Emergency and First Aid Procedures

## Section VII – Precautions for Safe Handling and Use

Steps to Be Taken in Case Material is Released or Spilled

Waste Disposal Method

Precautions to be Taken in Handling and Storing

Other Precautions

## Section VIII – Control Measures

Respiratory Protection (Specify Type)

| Ventilation | Local Exhaust | Special |
|---|---|---|
| | Mechanical (General) | Other |

| Protective Gloves | Eye Protection |
|---|---|

Other Protective Clothing or Equipment

Work/Hygienic Practices

Overall, it should be emphasized that the primary objective of touring the worksite is to evaluate the potential for adverse health effects upon all workers, in terms of hazards and ergonomic parameters. Hazards may be of physical, chemical, radiological, biological, or even of psychological origin. Ergonomic parameters relate not only to the physical condition of the workplace but also to *the manner in which and the tools with which* a given task is accomplished.

There are excellent references which address the many factors to be considered during a worksite walk-through;[15-17] and the types of hazards one may expect to find in various industries.[18] The approach here will be to present the most important factors in categorical fashion. In an effort to create a manageable and comprehensible format for the physician, both general and specific matters to consider during the walk-through are limited to twenty-three categories, not all of which will apply to every workplace.

### Walk-Through Imperatives*

- Search for all obvious and foreseeable *hazards* (VI-1; *p.77*) in the workplace. From the point of view of potential chemical exposure, Table 1 provides an excellent general overview of process-related sources of *chemical emissions* (VI-2; *p.77*). As part of this effort, the physician should compare observations made on the walk-through with data gathered during the pre-walk-through meeting.

- Observe the potential *biological availability* of all hazards. In other words, the physician should assess, for example, whether there is risk of physical contact following routine exposure to *skin* (VI-3; *p.89*), respiratory tract, G-I tract, eyes, and mucous membranes, or whether there is risk to any of these organ systems following foreseeable traumatic events.

---

* Refer to Appendix VI: *Walk-Through Imperatives Technical Guidelines* and the *Glossary* for greater details. A word or topic in this section for which there is a specific reference section in Appendix VI is indicated in parentheses, directly after the subject.

## TABLE I

## PERIODIC OR CONTINUOUS SOURCES
## OF CHEMICAL EMISSION (From 19)

The following unit or process operations are among those which may pose a risk of potential adverse chemical exposure to the worker:

**MATERIALS HANDLING** which gives rise to dust, gases, or vapors:

| | |
|---|---|
| Agitating | Pouring |
| Bagging | Recharging |
| Blending | Sampling |
| Debagging | Screening |
| Dismantling | Sieving |
| Emptying | Stirring |
| Exhausting | Sweeping |
| Grinding | Transferring |

**PROCESSES** causing emission of dust, vapors, particles, fumes, or gas:

| | |
|---|---|
| Abrading | Machining |
| Baking | Milling |
| Blasting | Painting |
| Coating | Planing |
| Crushing | Sanding |
| Curing | Sawing |
| Cutting | Soldering |
| Dipping | Spraying |
| Drilling | Welding |
| Drying | |

■ Evaluate potential *ergonomic problems* (VI-4; *p.89*).   The physician, for example, should observe whether the tools utilized or the manner in which the job is performed appear to pose a threat to the worker (especially lifting and materials handling).

Other important ergonomic factors to consider include:

- unnecessary or excessive repetitive motion

- condition and design of chairs and work stations

- routine job requirements which force the worker to assume *risky* non-anatomic positions (e.g., leaning or reaching to excess)

- design of tools and equipment

- work pace and scheduling of breaks and shifts

- any factors which may distract or prevent the worker from performing the job safely or accurately.

■ Evaluate the potential for *psychological stress* (VI-5; *p.90*) on the job.   This imperative is an elusive and difficult assignment for the physician.   In general, the physician must rely upon clinical training and professional instincts when tackling the problem of identifying psychological stress potential.   Factors to consider (some of which may overlap other walk-through imperative categories) include:

- workplace crowding

- presence (or absence) of personnel policies by which workers can complain or make suggestions

- adequacy of tools and physical surroundings relative to job assignment

- functional homeostasis of the workplace as gleaned from observations of interactions and communications among workers, management, and health and safety personnel, as well as measured by overall assessment of workplace productivity

- worker attitudes, especially during personal interview, as reflected by grievances, statements of satisfaction, or indifference.

When evaluating the potential for psychological stress during the walk-through, the physician may simply and wisely decide to refer all findings to an appropriate consultant, such as an industrial psychologist or human resources professional.

Walk-through information gathered in this context must be evaluated within the framework of written personnel policies, grievance procedures, employee satisfaction survey results, and similar evaluative tools, before determining data relevance and validity, much less before using such data to formulate recommendations. Despite these limitations, the physician should not be deterred from utilizing the walk-through as an opportunity for gathering information in this often overlooked hazard category.

**Inquire About and/or Evaluate:**

| | |
|---|---|
| Comfort | Temperature and relative humidity of the workplace (VI-6; *p.90*). |
| Facility Condition and Maintenance | Overall facility condition and maintenance with emphasis upon structural integrity, work surfaces, floors, stairs and ladders (VI-7; *p.92*). |
| Clothing and Personal Protection | Employee attire and personal protective equipment (e.g., hard hats, gloves, aprons, goggles, shoes, eyeglasses, ear plugs or muffs, and respirators), with attention to appropriateness to job requirements and working conditions (VI-8; *p.93*). |

| | |
|---|---|
| Lighting and Harmful Electromagnetic Frequencies | Lighting, both general and local (e.g., overhead and machine-mounted), and potential exposure to any harmful electromagnetic frequencies (e.g., U-V, lasers, RF, microwaves, etc.) (VI-9; *p.99*). |
| Ventilation | Ventilation, both general and local, with attention to effectiveness (e.g., a fan near a soldering bench may simply spread the fumes into a co-worker's breathing zone) (VI-10; *p.110*). |
| Housekeeping | Housekeeping (e.g., workplace neatness and organization) (VI-11; *p.114*). |
| Sanitation | Sanitation, eating, bathroom, locker, and shower facilities (VI-12; *p.115*). |
| Noise and Vibration | Background noise levels and potential vibration exposure (VI-13; *p.116*). |
| Signs and Labels | Adequacy and clarity of signage and labels of all kinds, including condition, contents, colors, and location (VI-14; *p.122*). |
| Chemical Containers | The presence and condition of all chemical containers (including compressed gases), with emphasis upon physical integrity, stability, and lid position (VI-15; *p.124*). |
| Chemical Storage | The condition, adequacy and location of chemical hazard storage facilities, with emphasis upon fire safety, chemical incompatibilities and the labeling of chemicals stored in bulk (e.g., sometimes chemical distribution spigots are positioned remote to bulk storage and are not labeled properly) (VI-16; *p.124*). |
| Confined Spaces | The location, designation, and accessibility of confined spaces (VI-17; *p.125*). |

| | |
|---|---|
| Emergency Equipment | The condition, location, and signage designation of emergency and safety equipment, including eyewash/shower units, first aid kits, fire extinguishers, self-contained breathing apparatus (SCBA) or other respirators, safety harnesses or life lines (VI-18; *p.127*). |
| First Aid Kits | Content and control of first aid kits, with emphasis upon worker access to potentially sedating medication such as pain relievers and cold preparations (VI-19; *p.127*). |
| Materials Handling | Degree and appropriateness of lifting and material handling requirements (VI-20; *p.129*). |
| Employee Behavior | The workplace behavior of employees, with specific attention to worker adherence to rules regarding use of safety equipment, eating, drinking, smoking, or engaging in any other form of prohibited or risk-taking behavior (VI-21; *p.129*). |
| Physician Self-Reporting | Any environmental condition which adversely impacts the physician personally, by producing irritation, or the development of local or systemic symptoms (but with the firm understanding that the human body is often a poor sensor of hazards present at harmful levels and may overreact to those present at acceptable levels) (VI-22; *p.129*). |
| Immediate Dangers to Life or Health | Categorical identification of serious, immediate dangers to life or health, requiring prompt notification of management and immediate mitigation (VI-23; *p.130*). |

Data gathered by the physician during the walk-through, in the context of assessing *program status*, is generally qualitative in nature.

Quantitative data, however, may be needed by the physician and should be requested from the employer to substantiate, to refute, or to clarify observations.   Sometimes, the employer already has such information and may have already provided it during the pre-walk-through meeting.   More often than not, however, such data does not exist and must be obtained from a qualified industrial hygienist.

*The Closing Meeting*:  In practice, the walk-through should be followed by a *closing meeting* during which data are organized and additional questions are asked.  During this meeting, the form and the extent of the final report or *deliverable* may be refined.  The physician, for example, may request that more information be provided before final conclusions can be drawn or recommendations can be made.  It is during this important meeting, moreover, that the physician may decide whether another visit is necessary.

The physician must remember that some observations and recommendations should be reported only verbally and preliminarily. Initial, hastily drawn conclusions and subsequently derived recommendations may be completely erroneous, inappropriate, or impractical.  Committing such conclusions and recommendations to writing will be a disservice to all parties and may preclude the physician's involvement in similar, future activities.  In other words, it is fine to say, *I think that the ventilation over by the vapor degreaser should be improved*;  but such a recommendation should not be committed to writing, until one knows that the actual measured ventilation is deficient and that the observed conditions during the walk-through are truly representative of a general condition of inadequate ventilation.   Finally, if committed to a written report, inappropriate, inadequate, or erroneous recommendations will also form an unfair *paper trail* with which the employer and, perhaps, even the physician must contend for years to come.

*Program Development*

Most of the information and guidelines discussed under the heading of *Program Status* can be utilized when the physician is asked to perform a walk-through for the purpose of *Program Development.* In the context of *Program Development,* as opposed to *Program Status,* however, it is even more important that as much data as possible be quantitative, rather than qualitative in nature. The reason for this assertion is based upon the final medical objective of the *Program Development* walk-through, namely, to synthesize the elements and the structure of a medical surveillance program. Since industrial hygiene factors usually *drive* the decision to perform medical surveillance, as a matter of routine procedure, the collection of quantitative data is particularly important. Naturally, as will be discussed, the physician cannot accomplish this task alone.

The methodology which must apply to the effort of *Program Development,* therefore, is to identify and, whenever possible, to quantify the potential exposure or dose of hazards to workers, on the basis of time, duration, location, organ system at risk, pathway of entry, job description, and unit operation.

Based upon the information gathered from the workplace, the medical surveillance program, in part, will be designed to answer the questions: *What medical tests should be administered to whom and when?* Before one has the ability to answer such important questions, it is necessary to gather the correct information and to organize it appropriately.

The methodology which describes the categorical data needed and the manner in which it should be organized is discussed in Appendix I to this volume. This methodology is called PATHMAX™ or *Parametric Approach Toward Health Maximization.* The development of PATHMAX™ requires a team effort in which the physician's walk-through assignment and analysis are only one part. Before proceeding, the reader is asked to review Appendix I.

It is important to re-emphasize that the physician cannot harvest all of the data necessary to complete a PATHMAX™ analysis during one or even several walk-throughs. The completion of a PATHMAX™ analysis requires a team effort, involving professionals from several disciplines, including personnel or human resources, safety, industrial hygiene, engineering, law, and medicine. The physician's job on the walk-through is to identify the qualitative nature and extent of hazards to workers in order to assist industrial hygiene and safety staff in their efforts to *measure* and *sample* those areas of greatest concern.

Ultimately, the physician should be responsible for developing and managing the information displayed in PATHMAX™ Matrix V which, in fact, tells the PATHMAX™ user *what medical surveillance tests should be performed on whom and when they should be performed.* Matrix V is the medical surveillance matrix. Its genesis is the natural consequence of a purposeful effort to utilize a broad spectrum of information, in conjunction with the application of corporate philosophy and priorities, in order to build a system of health risk management for new and current employees.

## Investigative Services

### Causality Determination

The physician may be requested to perform a workplace walk-through to gather the information necessary to answer a specific question or to solve a specific problem. One of the toughest of these problems is to learn whether a given medical complaint or finding is truly (probably) derived from or aggravated by conditions in the workplace. In occupational medical terms, the physician is asked to perform a *causality analysis.*

The logical skills and reasoning needed to embark upon this task are only infrequently discussed in the occupational medicine literature[20-25] and, regrettably, are not covered well in most of the standard occupational medical textbooks. As a result, there is little, if any, preparation for this task during specialty residency training and essentially none, whatsoever, in medical school. Despite these facts, the physician is still frequently called upon to perform a causality analysis, an assignment which can be an unforgiving challenge to even the most experienced and well-trained specialist. To the untrained, inexperienced or unmotivated physician, moreover, such a task can be both unpleasant and nearly impossible to accomplish. To make matters worse, an improperly performed causality analysis can inflict significant harm upon the patient, the employer, and the insurance carrier, as well as be a source of serious professional embarrassment and even potential liability[26] to the physician.

Appendix II discusses, in detail, the methodology, reasoning, and vocabulary which applies to the performance of a causality analysis within the worker compensation system in Colorado and in many other states within the United States. The important thing to remember is that Appendix II provides the logical framework within which any physician performing a walk-through can operate successfully. The reader is now referred to Appendix II before proceeding.

Recall from earlier discussions that one major purpose in performing the walk-through is to identify potential hazards to the worker. In the context of the causality analysis and the vocabulary of Appendix II, the purpose of the walk-through is to identify *probable cause or exposure*. Once identified, probable cause may be correlated with or excluded from a relationship with *probable effect or diagnosis,* assuming that the latter has been already established.

When evaluating the bases for probable exposure, the physician must visit the worker's environment, preferably when the worker is present and performing the routine duties of the job under representative conditions. If the worker is ill, on medical restrictions, or otherwise not available, the physician may have to observe another worker with job duties as similar as possible to those of the patient. The primary factors to be considered in evaluating the probability of *exposure* are:

- the presence, form, and *biological availability* of potential hazards

- confounding exposure factors in the workplace or in the patient's medical history which may account for other *exposure potential and experience*

- all *non-worker controlled factors* (e.g., employer choice of chemical, tool, or raw material, physical factors related to hazard confinement such as barriers, isolation, ventilation, etc.) which can either mitigate or aggravate hazard potential

- all *worker-controlled factors* (e.g., employee choice of chemical or tool, personal protective equipment, work practice, etc.) which can either mitigate or aggravate hazard potential

- the presence or absence of reports or *epidemiological evidence* of illness in co-workers with similar exposure potential.

Two major reasoning errors may be committed by failing to perform a walk-through as part of a *causality analysis* assignment. Each error is committed when the physician relies solely upon the patient's history when forming an *exposure or causal opinion*. The first is concluding that probable exposure or aggravation is present, when there is none (false positive exposure). The second is overlooking probable exposure or aggravation when, in fact, it exists (false negative exposure). Both errors can be ill-fated to the patient and to all related parties.

The *false positive exposure error* may lead the physician to declare the presence of probable workplace *disease* or aggravation when there is none. Under these circumstances, after the patient is first led to believe that the job alone is causing or aggravating illness, he or she may next be deprived of or be directed away from further medical evaluation.

The *false negative exposure error* is equally unfortunate and may lead to continuing injury and aggravation of what may be a totally preventable, job-related disorder.

Neither error should occur when the responsible physician performs a diligent workplace walk-through or engages in some other reliable form of environmental investigation.

Remember, almost without exception - patient history alone is simply not enough information with which one can formulate an exposure or causal opinion.  Corroboration by the employer, review of industrial hygiene data, and the walk-through may all be necessary before reaching final conclusions.

## *Risk Management or Cost Containment Activities*

The physician may occasionally be asked to perform a workplace walk-through as part of a broader, company-wide risk management or cost-containment program.  The circumstances under which this type of an assignment are varied but may be:

- a precursor to an *insurance inspection*

- prior to parent company or upper level *management visitation*

- part of an *occupational health acquisition audit* (i.e., either for the target company or for the buyer, prior to acquisition negotiations)

- part of a company-wide *illness and accident reduction program*, aimed at reducing health care and worker compensation costs.

If the physician elects to participate in these activities, most, if not all, of the previously discussed guidelines and recommendations can be utilized.  As a matter of practice, however, it is firmly recommended that the vast majority of physicians should avoid such assignments, unless they are very experienced and well-qualified.  Those specialists in the latter category should be certain to charge appropriately for their services because, like it or not, they will incur both direct, and indirect, potential liability when engaging in such activities.

## Regulatory or Audit Services

The physician is sometimes asked to perform a workplace walk-through to determine whether a client is in compliance with one or another municipal, state, or federal health and safety regulations.  The best advice for specialists and non-specialists alike is: *Don't do it!* Leave this job to the industrial hygiene and safety specialists or those insurance risk managers who have special training and experience in these matters.

Physicians must be alert for the situation in which one is performing a walk-through and is asked informally about regulatory compliance. This circumstance is no different from that which occurs when a patient who comes to see a cardiologist about chest pain inquires, *informally*, about a strange skin lesion on his neck.   Unless the physician is absolutely certain about the diagnosis, the patient should be referred to another specialist.   This same rule applies to the informal question regarding regulatory compliance.   Politely defer to someone better qualified, unless you wish to hear yourself quoted during governmental investigative or litigation hearings.

**Formulation of Accommodation Options for Disabled Workers**

*Background*

Historically, the occupational medicine physician has been called upon to tour the workplace in an effort to assist management in identifying appropriate accommodation options for disabled workers. After conducting a walk-through, the physician, for example, may recommend physical changes in the workplace, the purchase of new pieces of equipment, or the implementation of new or modified methods of performing a given job.

Given the advent of significant new federal legislation, the physician can expect to be called upon more frequently than ever to visit the workplace to assist in formulating accommodation options for disabled workers.

*The Americans with Disabilities Act (ADA or PL101-336)[1]*

The Americans with Disabilities Act (ADA) will become effective on July 26, 1992.  The provisions and details of this act are enormously complex and far reaching, and are, thus, well beyond the scope of this text.

One major feature and consequence of the ADA, however, which is relevant and deserves special attention here, is that the law provides the opportunity for disabled* workers to engage in job activities from which they may have been historically excluded.

Prior to ADA, exclusion was not necessarily based upon the worker's inability to perform the job, but instead, may have been based upon the employer's unwillingness to provide *reasonable accommodations* to the worker.   ADA, in essence, prohibits, under federal law, any employer from discriminating against a disabled worker who is otherwise qualified to perform the *essential functions of the job*. ADA also specifies that the employer will provide reasonable accommodations to the qualified disabled worker, as part of the employer's ADA compliance responsibilities.

*Within the era of ADA*, it is inevitable that employers will have more questions than ever regarding:

- What is a reasonable accommodation for a given disabled worker?

- How should an accommodation be implemented?

- How should an accommodation be maintained?

- How should an accommodation change when the job requirements change or, just as importantly, when the disability or health status of the employee changes?

- What are the *essential functions* of the job?

---

\* Under ADA disability means:
   **A physical or mental impairment that substantially limits one or more of the major life activities... A record of such impairment...or being regarded as having such impairment...**[1]

The precise answers to these questions will require the physician to be more knowledgeable than ever about the details of the employee's disability, the conditions in the workplace, and the precise requirements of each job which the employee is expected to perform.

When evaluating the needs of a particular disabled worker, there may be several accommodation options open to the employer, evident to the physician, or even suggested by the disabled employee. Table II provides the reader with a number of accommodation options with which the physician should be familiar.

The options are divided into the following basic categories:

- Structural or mechanical

- Procedural or methodological

- Expectational or functional

- Environmental

- Organizational

Within each category just a few basic examples are provided to guide the physician regarding specific options. Dozens of additional options could be added to each category; but, the list provided in Table II should assist the physician under most circumstances, when he or she engages the challenging task of recommending the correct *accommodation prescription*.

## FINANCIAL CONSIDERATIONS

The physician is usually asked to perform a walk-through for well-defined reasons. In the United States, often the physician is paid for this service. Sometimes, however, the overall task of familiarizing oneself with the workplace is provided without fee, as long as the

# TABLE II
## SELECTED ADA ACCOMMODATION OPTIONS

STRUCTURAL OR MECHANICAL

Mobility-Related
  Access
  Ramps
  Handles
  Elevators
  Automobile
  Lifts
Work Station-Related
  Chairs
  Foot rests
  Table design
  Lighting
Architectural
  Sanitary facilities
  Trailers
  Enclosure or isolation
  Home-based work

PROCEDURAL OR METHODOLOGICAL
  (e.g. Job Performance Modification)
Tools
Order of Operations
Job Sharing
Modification of Shift or Working
  Hours
Task Sharing
  Reading assistance to visually
    impaired
  Sign language assistance to hearing
    impaired
Modification of Personal Protective
  Equipment - Change in:
  Respirator type
  Goggles
  Gloves
  Clothing
  Ear plugs or muffs

EXPECTATIONAL OR FUNCTIONAL
  (e.g. Job Productivity Modification)
Rate of Production (piecework)
Amount (quotas)
Quality of Product
Qualifications of Employee

ENVIRONMENTAL
Physical
Modification of:
  Temperature or Relative Humidity
  Ventilation
  Ionizing Radiation
  Electromagnetic Radiation
    UV
    Visible
  Noise
  Vibration
Chemical or Biological*
Accommodations for persons who are:
  Hypersensitive-allergic
  Hypersusceptible or predisposed
Psychological
  Modification of:
    Stress
      Peer-related
      Management-related
    Crowding

ORGANIZATIONAL
  Conflict Management
  Conflict Resolution
    Mediation
    Arbitration
    Employee Assistance

---

* Mitigation options include: isolation, dilution (ventilation), elimination, substitution, and personal protection.

effort is part of a larger, funded assignment (e.g., a retainer for provision of clinical or consulting services or for performance of routine duties within a corporate medical job position). Overall, however, when appropriate, the physician should charge a fee for evaluating the workplace.

The reasons for the recommendation to charge a fee for performing a workplace walk-through are fundamentally ingrained within the framework of preventive medicine. One primary objective of this medical discipline is to *prevent* disease and injury. The *treatment* of disease and injury, however, usually represents a large portion of the occupational medical revenues earned by the medical and hospital-based community. If, in the process of performing a walk-through, the physician is successful in reducing occupational morbidity, practice revenues for treating accidents and illnesses will, hopefully, decline. As a logical economic sequitur, therefore, the occupational medicine physician or one who is acting in such a capacity should replace previously earned *morbidity - related* income with income earned from *preventive activities*.

Charging for preventive walk-through services, without question, is a legitimate exercise, directed toward achieving the goal of health resources reallocation, which is entirely compatible with the basic tenets and objectives of preventive medicine. Clearly, moreover, charging for these services forces the client and the physician to take the walk-through exercise more seriously.

## SUMMARY

The workplace walk-through is an essential activity for any physician who practices occupational medicine. Without the information gathered during the walk-through, probable cause or potential exposure is difficult, if not impossible to determine. The physician must perform this task personally and not delegate it to others*, because the information gathered on the walk-through provides the foundation upon which the rational exercise of *causality determination* is built. Walk-through data, moreover, can be utilized within an integrated system of data management from which appropriate recommendations for occupational medical surveillance can be derived.

The invitation to perform a workplace walk-through may come from many sources and under a wide variety of circumstances. The tasks requested of the physician conducting the walk-through are varied; and some should not be undertaken, unless the physician is specially trained or experienced. In most cases, however, the diligent and concerned physician can complete the walk-through without problems and can, therefore, better serve the needs of both patients and employers.

---

* When it is not possible for the physician to visit the workplace or when the physician is not invited, it is reasonable to assign the task to a qualified industrial hygienist or another technical person with whom the physician may collaborate and upon whose report the physician may depend.

# APPENDIX I

## PATHMAX™
### (Parametric Approach Toward Health Maximization)

## Introduction

PATHMAX™ is an occupational health paradigm within which many functions of the respective occupational health team members are unified. Included within this methodology are the activities of the industrial safety engineer, the industrial hygienist, the occupational health legal advisor, and the occupational health physician and nurse.

PATHMAX™ is a series of five operational matrices which, when viewed together, demonstrate how critical information from environmental characterization efforts is tied directly to medical surveillance recommendations. When utilized separately, each PATHMAX™ matrix provides valuable information regarding potential risk and/or recommendations to control risk. PATHMAX™ is not derived from nor is it in any way dependent upon any of the numerous occupational health information systems developed over the past two decades by various corporations,[27-29] universities, governmental agencies,[30] insurance companies, and private consultants.

## Structure and Purpose

The PATHMAX™ system consists of five two-dimensional matrices, each of which represents various data at a specific point in time. Theoretically, each matrix can be expanded into a third or time dimension, if one wishes to examine the historical aspects or to contemplate the prognostic features of the data base.

PATHMAX™ Matrix I is derived from the efforts of the medical, industrial engineering, industrial hygiene, and safety professionals assigned to the PATHMAX™ team. The elements of PATHMAX™ Matrix II are provided through the efforts of personnel or human resources professionals who are familiar with job descriptions and plant unit operations. PATHMAX™ Matrix III is derived parametrically from Matrices I and II by computer analysts, programmers, or other technical members of the PATHMAX™ team. The construction of PATHMAX™ Matrix IV reflects the efforts of legal, medical, and management professionals. Finally, the parametric transformation of Matrices III and IV into Matrix V is performed by those same professionals who derived Matrix III. Matrix V, the medical surveillance matrix, is actually only the core matrix of a more elaborate system which instructs the medical practitioner regarding: *What tests should be performed upon which workers and when each should be conducted.* The complementary system which illustrates how to organize or to massage the information collected from Matrix V, and then, how to make decisions about worker fitness, job restrictions, and job modification, based upon that information, is not discussed in this appendix.

## Benefits

The five PATHMAX™ matrices are intended to illustrate a large quantity of information, some of which may not be initially available. PATHMAX™ methodology, nevertheless, can be implemented, in general terms, without knowing the details of every element in each matrix. In this sense, PATHMAX™ is a logical didactic construct with which the occupational health professional and others can better understand the manner in which the status of the environment and the medical surveillance requirements are integrated.

It is important to re-emphasize that each PATHMAX™ matrix is a *snapshot* in time, which characterizes the temporary status of knowledge regarding the factors displayed. As information increases or decreases, it becomes a simple matter to discern whether matrix elements are full or empty. In other words, it is easy to see *where the holes are.* From this point of view, the PATHMAX™ methodology represents an excellent risk management tool with which non-technical managers can functionally interact with their technical support staff to assign health and safety priorities.

The versatility of PATHMAX™ also permits the user to reconstruct any given matrix at any previous point in time in an effort to compare *what was* to *what should have been*, the state of knowledge regarding health, safety, human resources, and legal requirements in the past. This reconstruction exercise permits one, inferentially, to evaluate the history of any occupational safety and health program, sometimes, in an effort to develop a prognosis for predicting the future incidence of occupational diseases and costs of health care. Naturally, this exercise also permits comparison of historical programs with contemporary ones.

*EXAMPLE: The Acquisition Audit*

The utility of PATHMAX™ methodology as part of an acquisition audit, for example, can be easily illustrated. If company *A* wishes to purchase company *B*, *A*, of course, will engage in a comprehensive audit of *B*'s financial history and make predictions about *B*'s viability in future markets. In recent years, it has become more common for *A*'s audit teams also to include an assessment of environmental, occupational, safety, and health liability as part of the overall acquisition audit effort. If *B* (the acquisition target), for example, is a chemical manufacturing company, it will interest *A* to know whether *B* has historically maintained a state-of-the-art occupational safety and health program.

By reconstructing historical versions of PATHMAX™ Matrices I and III, *A*'s auditors may discover, for example, that between 1963 and 1972, *B* used large amounts of benzene (a human leukemogen) and learn further that *B*'s actual surveillance programs over those years failed to recognize or to monitor this hazard. Under these conditions, after *A*'s consultants have developed inferences about the targets and extent of benzene exposure and have projected disease incidence, in conjunction with consideration of disease latency period, they can calculate the estimated costs of future benzene-related morbidity and mortality. These cost predictions can, then, be factored into *A*'s financial negotiations with *B* prior to acquisition.

This previous example illustrates only one of many special uses of the PATHMAX™ system for organizing and then analyzing important occupational safety and health information.

Another major use of the PATHMAX™ system is for the purpose of training and educating a whole array of individuals including medical students, residents, physicians, nurses, industrial hygienists, safety personnel, and managers.  No other logical system demonstrates both the derivational and functional interdependency of information gathered from such varied but familiar sources.  The PATHMAX™ system, therefore, is the ideal training tool to build the cohesion necessary to optimize the performance of the occupational safety and health team, by showing the student the need for and the mechanism by which information must be shared.

**Approach to Matrix Construction**

In the accompanying five figures, Matrices I-V will be discussed individually.  The matrices will be followed by a working example of Matrix V.

First, consider a few basic rules.  Each matrix element should contain information which is as contemporary as possible with information in other elements.  When beginning, any qualitative or quantitative information may be useful in element specification for Matrices I-III.  Whenever possible, as the PATHMAX™ effort matures, information in these matrices should become as quantitative as possible. Information in the elements of Matrices IV and V will almost always be qualitative.  Information for Matrices I-III can be gathered to some extent from the pre-walk-through meeting and during the walk-through itself.

The five PATHMAX™ matrices will be generally discussed in the context of a Hazardous Waste Site Operation.  After this discussion and presentation, a fully functional Matrix V will be presented with four important *back-up* or conditional matrices.

*The reader is alerted that these examples are for didactic purposes only and are not presented as recommendations. None of the information in any of the matrices should be utilized without expert consultation.*

## FIGURE I-1
## MATRIX I – STATIC HAZARD ANALYSIS MATRIX

Potential Hazards or Stressors as a Function of Selected Unit Operations

POTENTIAL HAZARDS

| UNIT OPERATIONS | Chemical | Physical | Radiological | Biological | etc..... |
|---|---|---|---|---|---|
| Sample procurement | | | | | |
| Storage | | | | | |
| Incineration | | | | | |
| Sample analysis etc. | | | | | |
| etc..... | | | | | |

*Derivation:*  Primarily the efforts of the industrial hygienist, safety, and industrial engineers.

*Purpose:*  Characterization of the workplace by unit operations, each operation of which is fixed, physically in location and is potentially unique in terms of risk.

*Matrix Element Information:*  May vary greatly from categorical (yes or no) to semi or fully quantitative data, including, for example, percentage of threshold limit values and/or actual environmental monitoring data. Records of accidental, episodic exposures and potential pathways of entry or injury may be included.

*Limitation:*  Matrix I may not apply to smaller companies in which unit operations are not discernible. Even in larger companies or more complex workplaces, unit operations may not be unique.

*Comments:*  The columns should list individual, specific, potential (biologically available) hazards, although this figure illustrates the hazards only categorically.

## FIGURE I-2
## MATRIX II – THE DYNAMIC ACTIVITIES MATRIX

Activities of Selected Site Personnel in Relationship to Selected Unit Operations

### UNIT OPERATIONS

|  | Sample procurement | Storage | Incineration | Sample analysis | etc..... |
|---|---|---|---|---|---|
| Geohydrologist |  |  |  |  |  |
| Incinerator Operator |  |  |  |  |  |
| Safety/Health Officer |  |  |  |  |  |
| Bulldozer Operator |  |  |  |  |  |
| Laboratory Technician |  |  |  |  |  |
| Maintenance |  |  |  |  |  |
| etc..... |  |  |  |  |  |

JOB DESCRIPTION

*Derivation:* Primarily from the efforts of human resource personnel, industrial engineers, and from interviews with individual workers.

*Purpose:* Identification of job descriptions in terms of encounter with or time spent in various unit operations.

*Matrix Element Information:* May vary greatly from categorical (yes or no) to semi or fully quantitative data including, for example, actual hours or percentage time spent within each unit operation.

*Limitation:* Matrix II may not apply to smaller companies or to larger ones in which job descriptions are poorly characterized or may change too quickly over time.

*Comments:* Ideally, information should be sought regarding the actual amount of time each worker spends within each unit operation. Such data are useful in specifying the actual dose potential of a given hazard to the worker.

## FIGURE I-3
## MATRIX III– THE BIOLOGICAL AVAILABILITY MATRIX

Potential Hazards or Stressors as a Function of Selected Job Categories

POTENTIAL HAZARD

| JOB DESCRIPTION | Chemical | Physical | Radiological | Biological | etc..... |
|---|---|---|---|---|---|
| Geohydrologist | | | | | |
| Incinerator Operator | | | | | |
| Safety/Health Officer | | | | | |
| Bulldozer Operator | | | | | |
| Laboratory Technician | | | | | |
| Maintenance | | | | | |
| etc..... | | | | | |

*Derivation:* Using elementary parametric notions [e.g., $X = f_1(t)$ and $Y = f_2(t)$, then derive $Y = f_3(X)$]. Matrix III can be derived from Matrices I and II. This parametric transformation is performed by computer analysts or other technical personnel.

*Purpose:* Identification of the potential (biologically available) hazards which may be encountered by workers operating within each definable job description.

*Matrix Element Information:* Same format as Matrix I.

*Limitations:* For smaller companies in which the workplace is small and job descriptions are not well defined, the mere presence of the worker in the workplace may place him or her at risk for exposure to most or all potential hazards. Even in larger workplaces, job descriptions may be poorly defined; and the rows of Matrix III may need to specify more general categories such as *clerical, in-shop, degreaser area, executive, carpool,* etc.

*Comments:* Matrix III is the usual starting point for applying PATHMAX™ methodology to smaller workplaces. The columns should list individual, specific, potential (biologically available) hazards, although this figure illustrates the hazards only categorically.

## FIGURE I-4
## MATRIX IV – THE REGULATORY HEALTH RISK MANAGEMENT MATRIX

Individual Health Surveillance Parameters Necessary to Monitor Potential Hazards or Stressors

POTENTIAL HAZARD

|  | Chemical | Physical | Radiological | Biological | etc..... |
|---|---|---|---|---|---|
| Core History & Physical Examination |  |  |  |  |  |
| Heavy Metal Screen |  |  |  |  |  |
| Audio-screen |  |  |  |  |  |
| EKG |  |  |  |  |  |
| etc..... |  |  |  |  |  |

HEALTH SURVEILLANCE PARAMETERS

*Derivation:* The efforts of attorneys, medical personnel, and management. Information is derived from existing laws, governmental or industry-wide recommendations, corporate philosophical and prioritization guidelines, medical judgment, and budgetary considerations.

*Purpose:* Consolidation of information regarding what tests or procedures are needed for surveying a population potentially exposed to specific hazards.

*Matrix Element Information:* May contain *yes*, *no*, or *conditional* options. The *conditional* option may specify that certain tests are to be performed only under certain exposure conditions, at certain exposure levels (e.g., above the Action Level), or for specific individuals (e.g., those who are pregnant, asthmatic, etc.). Complex *conditional* options are specified by using appropriate footnotes.

*Limitations:* Because of the rapidly changing scientific, political, and legal climate of occupational medicine, it is sometimes difficult to decide whether to recommend a given test or procedure, especially if the cost-effectiveness of such a decision is uncertain. This matrix, therefore, may not always reflect *state-of-the-art* decision-making, especially as the cost of testing becomes burdensome.

*Comments:* Despite the notion that health surveillance *go/no go* decisions are based upon the level of potential exposure, many clients say *go* simply based upon the categorical presence of potential hazard to the worker.

Matrix IV delineates those health surveillance parameters which have been specified to monitor each potential hazard category. Derivation of the elements of Matrix IV comes from experience, medical and legal research, client recommendations, and the *state-of-the-art* practice of occupational medicine.

## FIGURE I-5

## MATRIX V – THE HEALTH SURVEILLANCE IMPLEMENTATION MATRIX

Health Surveillance Parameters as a Function of Each Job Category

JOB DESCRIPTION

|  | Geohydrol-ogist | | | Incinerator Operator | | | Safety & Health Off. | | | Bulldozer Operator | | | Lab Technician | | | Mainten-ance | | | etc...... | | |
|---|---|---|---|---|---|---|---|---|---|---|---|---|---|---|---|---|---|---|---|---|---|
| HEALTH PARAMETERS | PP | INT | TER | PP | INT | TER | PP | INT | TER | PP | INT | TER | PP | INT | TER | PP | INT | TER | PP | INT | TER |
| Core History & Phys. exam | | | | | | | | | | | | | | | | | | | | | |
| Heavy Metal Screen | | | | | | | | | | | | | | | | | | | | | |
| Audio-screen | | | | | | | | | | | | | | | | | | | | | |
| EKG | | | | | | | | | | | | | | | | | | | | | |
| etc...... | | | | | | | | | | | | | | | | | | | | | |

PP = preplacement; INT = Interval; TER = termination

*Derivation:* Same as the derivation of Matrix III, except that Matrix V is derived from Matrices III and IV. It is always customized by adding one or more sub-columns within each *Job Description* column, each of which specifies an employment timeframe within which surveillance tests are to be conducted (e.g., PP = preplacement; INT = interval; TER = termination.).

*Purpose:* Matrix V is the core instrument for characterizing the health surveillance program. It specifies *What to do to whom and when* - or, namely, what medical procedures, tests, or protocols are to be administered to a given individual at any given point in time.

*Matrix Element Information:* (also see Matrix IV) - Elements within Matrix V may include one or more of an array of responses. These are *yes, no, conditional, DBRP* or *Determined by Responsible Physician,* or a numbered footnote. When the conditional response is present, the user is referred to a *conditional* matrix (see Figures I-7 to I-10) from which the final instruction is obtained. This matrix will contain the same column categories as Matrix V. Row categories, however, do not include health surveillance parameters. Instead, they include *conditional situations.* For example, if the user of Matrix V wants to know whether to obtain a back x-ray from a bulldozer operator at the pre-placement exam, he will be referred to a conditional matrix for back x-rays (Figure I-7). This matrix will have the same columns as Matrix V, but the row will list various situations, such as *Higher Risk and Back X-Rays within 6 Months but No New Symptoms,* or *Average Risk and Baseline Back X-Ray Exists and is Available.* Elements within the conditional matrix will then provide *yes, no, DBRP,* or numerical footnotes to aid in the decision-making process. *DBRP* or *Determined by Responsible Physician* is included as an option within Matrix V and conditional matrices because certain *go/no go* decisions cannot be prescribed, in advance, for all situations and, therefore, must be made by the physician who is actually conducting the examination. Corporate guidelines, for example, may have dictated from Matrix IV (via a *conditional* footnote) that the *Responsible Physician* may order thyroid function testing, at the time of the physical examination, on individuals who present with certain complaints and/or physical findings. Likewise, the decision to perform a rectal or breast exam may be *DBRP*, depending upon the nature, family history, physical findings, or other factors which only the physician will know at the time the decision is made. Finally, numerical footnotes are often needed to specify information too extensive to be written into the limited space within the matrix element.

*Limitations:* Despite its basic simplicity, those who administer Matrix V will require orientation and training. The construction of Matrix V can be costly, if there are many job descriptions to be considered. Costs of construction vary from $250 - $10,000+, depending upon the complexity of the workplace.

*Comments:* Additional customization of Matrix V occurs when client philosophy or priorities creates the addition of extra rows in the matrix. These rows contain medical surveillance parameters which were not derived from Matrix IV analysis. These additions, however, are usually straightforward and easy to understand. For example, a cardiac stress test may be included and made available to those in the executive job category. Likewise, *substance abuse screening* or *drug testing* may be included for all job categories. These customized entries are usually not added for regulatory reasons. Sometimes, for example, they are established, by employer request, to render PATHMAX™ Matrix V compatible with less well-organized, *pre-PATHMAX*™, health and safety program manuals.

OPERATIONAL EXAMPLE OF PATHMAX™ MATRIX V WITH
SUPPORTING CONDITIONAL MATRICES

Examples of Matrix V and four conditional matrices (Back X-Ray, Chest X-Ray, Pulmonary Function Test, and Resting EKG) are presented in Figures I-6 through I-10. These matrices were constructed as part of the PATHMAX™ Analysis for a Hazardous Waste Site Operation, involving five basic job categories: Executive Management (including accounting and marketing), Plant Operators, Landfill Operators, Lab/Environmental Health Science, and Maintenance. The four conditional, supporting matrices were chosen for those surveillance elements which are costly and involve *invasive* modalities (e.g., x-rays) and for those elements for which more complex *go/no go* decision-making is required. The directives in these matrices should be self-evident, but as a matter of practice, physicians who implement this matrix will require at least 3-4 hours of training and will have access to an occupational medicine specialist for questions during surveillance implementation.

## FIGURE I–6 – NOTES
## MATRIX V – EXAMPLE

### Legend

DBRP...Determined by Responsible Physician
C.........Conditional
PP........Preplacement
INT......Interval
TER.....Termination

### Footnotes

(1)....Yes, if over age 45 and has not had exam within one year, otherwise DBRP
(2)....Refer to CXR for routinely exposed–preplacement
(3)....Refer to BXR for routinely exposed–preplacement
(4)....Refer to EKG for routinely exposed–preplacement
(5)....DBRP, but exam should be strongly considered
(6).... > or = age 40: annual unless acutely exposed; < age 40: biannual unless acutely exposed
(7)....Interval history will be obtained
(8)....Annual unless acutely exposed
(9)....More frequent exams may be required if environmental exposure data or potential exposure warrants increased frequency
(10)...Refer to CXR for routinely exposed–interval
(11)...Refer to BXR for routinely exposed–interval
(12)...Refer to EKG for routinely exposed–interval
(13)...Refer to CXR for routinely exposed–termination
(14)...Refer to BXR for routinely exposed–termination
(15)...Refer to EKG for routinely exposed–termination
(16)...Refer to PFT for Executive examinations at relevant timeframe.
(17)...Refer to CXR for Executive examinations – preplacement.
(18)...Refer ro CXR for Executive examinations – interval.
(19)...Refer to CXR for Executive examinations – termination.
(20)...Refer to BXR for Executive examinations – preplacement.
(21)...Refer to BXR for Executive examinations at relevant timeframe.


*......Freeze Serum from (2) Red Top Vacutainers for later use, if necessary up to 6 months
**.....From Urine Obtain: As, Hg, Cd; From Serum Obtain: Pb, ZPP, (Optional–Cr (VI), Ni)

# FIGURE I-6
## MATRIX V – EXAMPLE
(For didactic purposes only. Do not utilize any of the matrix information without expert consultation)

| DESCRIPTION OF EVALUATION PARAMETERS | EXECUTIVE MANAGEMENT incl ACCOUNTING and MARKETING | | | PLANT OPERATORS | | | LANDFILL OPERATORS | | | LAB/EHS | | | MAINTENANCE | | |
|---|---|---|---|---|---|---|---|---|---|---|---|---|---|---|---|
| | PP | INT(6) | TER | PP | INT(8,9) | TER | PP | INT(8,9) | TER | PP | INT(8,9) | TER | PP | INT(8,9) | TER |
| Core Occupational History | Yes | Yes (7) | Yes (7) | Yes | Yes (7) | Yes (7) | Yes | Yes (7) | Yes (7) | Yes | Yes (7) | Yes (7) | Yes | Yes (7) | Yes (7) |
| Core Medical/Surgical History | Yes | Yes (7) | Yes (7) | Yes | Yes (7) | Yes (7) | Yes | Yes (7) | Yes (7) | Yes | Yes (7) | Yes (7) | Yes | Yes (7) | Yes (7) |
| Complete Physical Exam. | Yes | Yes | Yes | Yes | Yes | Yes | Yes | Yes | Yes | Yes | Yes | Yes | Yes | Yes | Yes |
| Complete Blood Count with Diff. | Yes | Yes | Yes | Yes | Yes | Yes | Yes | Yes | Yes | Yes | Yes | Yes | Yes | Yes | Yes |
| Urinalysis & Micro. Exam. | Yes | Yes | Yes | Yes | Yes | Yes | Yes | Yes | Yes | Yes | Yes | Yes | Yes | Yes | Yes |
| SMA-25* | Yes | Yes | Yes | Yes | Yes | Yes | Yes | Yes | Yes | Yes | Yes | Yes | Yes | Yes | Yes |
| Thyroid Screen | Yes | No | Yes | Yes | Yes | Yes | Yes | DBRP | Yes | Yes | Yes | Yes | Yes | Yes | Yes |
| Risk Factor Profile | Yes | DBRP | Yes | DBRP | DBRP | DBRP | DBRP | DBRP | DBRP | DBRP | DBRP | DBRP | DBRP | DBRP | DBRP |
| Resting EKG | Yes | Yes | Yes | Yes | Yes | Yes | Yes | C (12) | C (15) | Yes | Yes | No | Yes | Yes | Yes |
| Cardiac Treadmill | DBRP | DBRP | DBRP | DBRP | DBRP | DBRP | No | No | No | DBRP | No | No | DBRP | DBRP | DBRP |
| Pulmonary Function Exam. | C (16) | C (16) | C (16) | Yes | Yes | Yes | Yes | Yes | Yes | Yes | Yes | Yes | Yes | Yes | Yes |
| Vision Test | Yes | Yes | Yes | Yes | Yes | Yes | Yes | Yes | Yes | Yes | Yes | Yes | Yes | Yes | Yes |
| Tonometry | DBRP | DBRP | DBRP | Yes | Yes | Yes | Yes | No | No | Yes | Yes | Yes | Yes | Yes | Yes |
| Procto Exam. | No | No | No | Yes (1) | DBRP | Yes (1) | No | No | No | Yes (1) | DBRP | Yes (1) | Yes (1) | DBRP | Yes (1) |
| Audiometry | Yes | DBRP | Yes | Yes | Yes | Yes | Yes | Yes | Yes | Yes | DBRP | Yes | Yes | Yes | Yes |
| Chest X-Ray | C (17) | C (18) | C (19) | C (2) | C (10) | C (13) | C (2) | C (10) | C (13) | C (2) | C (10) | C (13) | C (2) | C (10) | C (13) |
| Back X-Ray | C (20) | C (21) | C (21) | C (3) | C (11) | C (14) | C (3) | C (11) | C (14) | C (3) | C (11) | C (14) | C (3) | C (11) | C (14) |
| Heavy Metal Screen** | DBRP | No | No | Yes | Yes | Yes | Yes | Yes | Yes | Yes | Yes | Yes | Yes | Yes | Yes |
| Cholinesterase Screen | DBRP | No | No | Yes | Yes | Yes | Yes | DBRP | Yes | Yes | Yes | Yes | Yes | Yes | Yes |
| Creatinine Clearance | DBRP | No | No | Yes | Yes | Yes | Yes | Yes | Yes | Yes | Yes | Yes | Yes | Yes | Yes |
| Substance Abuse Exam. | Yes | Yes | Yes | Yes | Yes | Yes | Yes | Yes | Yes | Yes | Yes | Yes | Yes | Yes | Yes |
| Neuropsychiatric Eval. | DBRP | No | DBRP | DBRP | DBRP | DBRP | DBRP | DBRP | DBRP | DBRP (5) | DBRP | DBRP (5) | DBRP (5) | DBRP | DBRP (5) |
| Serum PCB | DBRP | No | No | Yes | Yes | Yes | DBRP | DBRP | DBRP | Yes | Yes | Yes | Yes | Yes | Yes |
| ILO-UICC "B" Reading | DBRP | No | No | Yes | No | Yes | Yes | Yes | Yes | Yes | DBRP | Yes | Yes | Yes | Yes |

FIGURE I-7

**CONDITIONAL MATRIX**
**CRITERIA FOR BACK X-RAY  ***

| DESCRIPTION OF EXAMINATION | EXECUTIVE EXAMINATIONS | | | ROUTINELY EXPOSED | | | NON-EXPOSED | | | SPECIAL EXPOSURE | | |
|---|---|---|---|---|---|---|---|---|---|---|---|---|
| | PP | INT | TER | PP | INT | TER | PP | INT | TER | PP | INT | TER |
| ** Higher Risk and Back X–Rays Within 6 months but No Symptoms — All Ages | N | | N | DBRP | N | DBRP | N | N | N | DBRP | N | DBRP |
| ** Higher Risk and Back X–Rays Within 6 months — All Ages | Y | DBRP | DBRP | Y | DBRP | Y | Y | DBRP | Y | Y | DBRP | Y |
| *** Higher Risk and No Back X–Ray Within 6 months — All Ages | Y | DBRP | DBRP | Y | DBRP | Y | Y | DBRP | Y | Y | DBRP | Y |
| *** Average Risk and Baseline Back X–Ray Exists and Is Available — All Ages | N | N | N | N | N | N | N | N | N | N | N | N |
| *** Average Risk and No X–Ray of Back Has Ever Been Taken — All Ages | DBRP | DBRP | DBRP | DBRP | DBRP | DBRP | DBRP | DBRP | DBRP | DBRP | DBRP | DBRP |

FIGURE I-7 - NOTES
CONDITIONAL MATRIX
CRITERIA FOR BACK X-RAY

* Under all conditions, no female employee shall undergo an X-ray unless she certifies in writing that she is in the process of menses, is taking birth control medication, or that she is not pregnant.

** Higher Risk – (includes one or more of the factors listed below)
- Weight more than 1.2 x ideal weight
- Previous abnormal back X-ray spondylolysis, spondylolisthesis, fracture
- History of previous serious back injury, surgery, congenital problems or disease processes which might involve the back
- Symptomatic

*** Average Risk – (all factors must apply)
- Weight equal to or less than 1.2 x ideal weight
- No history of abnormal back X-ray or other problems
- No symptoms

Note: If all Average Risk factors apply and one or more Higher Risk factors apply, the patient should be considered Higher Risk

PP = Pre-Placement evaluation
INT = Interval evaluation
TER = Termination evaluation

DBRP = Determined by Responsible Physician
Y  = Yes
N  = No

FIGURE I-8
CONDITIONAL MATRIX
CRITERIA FOR CHEST X-RAY*

| DESCRIPTION OF EXAMINATION | EXECUTIVE EXAMINATIONS | | | ROUTINELY EXPOSED | | | NON-EXPOSED | | | SPECIAL EXPOSURE | | |
|---|---|---|---|---|---|---|---|---|---|---|---|---|
| | PP | INT | TER | PP | INT | TER | PP | INT | TER | PP | INT | TER |
| **Smoker** or Former Smoker*** and No Chest X-ray within 3 months, or cannot provide recent chest X-ray:** | | | | | | | | | | | | |
| Age: Under 40 | Y | DBRP | Y | Y | Y | Y | Y | DBRP | DBRP | Y | Y | Y |
| Age: 40 or over | Y | Y | Y | Y | Y | Y | Y | DBRP | DBRP | Y | Y | Y |
| **Smoker** or Former Smoker*** and Chest X-ray within 3 months, and can provide:** | | | | | | | | | | | | |
| Age: All Ages | N | N | N | DBRP | DBRP | DBRP | N | N | N | DBRP | DBRP | DBRP |
| **Nonsmoker without Symptoms or Findings, and No Chest X-ray within 6 months or cannot provide recent Chest X-ray:** | | | | | | | | | | | | |
| Age: All Ages | Y | DBRP | DBRP | Y | Y | Y | Y | DBRP | DBRP | Y | Y | Y |
| **Nonsmoker without Symptoms or Findings, and Chest X-ray within 6 months, and can provide:** | | | | | | | | | | | | |
| Age: All Ages | Y | N | N | Y | N | Y | DBRP | N | N | Y | N | Y |
| **Nonsmoker with Symptoms and /or Findings** | | | | | | | | | | | | |
| Age: All Ages | Y | DBRP | Y | Y | Y | Y | Y | DBRP | Y | Y | Y | Y |

FIGURE 1-8 –NOTES
CONDITIONAL MATRIX
CRITERIA FOR CHEST X-RAY

* Under all conditions, no female employee shall undergo an X-ray unless she certifies in writing that she is in the process of menses, is taking birth control medication, or that she is not pregnant.

** Smoker: current routine use of at least one cigarette/day, or one cigar/day, or one pipeful/day, or one joint (marijuana)/day.

*** Former Smoker: History of at least one cigarette/day, or one cigar/day, or one pipeful/day, or one joint (marijuana)/day on a routine basis for one year or more.

PP  = Pre-Placement evaluation
INT = Interval evaluation
TER = Termination evaluation

DBRP = Determined by Responsible Physician
Y   = Yes
N   = No

FIGURE I-9

**CONDITIONAL MATRIX**
**CRITERIA FOR PULMONARY FUNCTION TEST**

| DESCRIPTION OF EXAMINATION | EXECUTIVE EXAMINATIONS | | | ROUTINELY EXPOSED | | | NON-EXPOSED | | | SPECIAL EXPOSURE | | |
|---|---|---|---|---|---|---|---|---|---|---|---|---|
| | PP | INT | TER | PP | INT | TER | PP | INT | TER | PP | INT | TER |
| * Higher Risk All Ages | Y | Y | Y | Y | Y | Y | Y | DBRP | DBRP | Y | Y | Y |
| ** Average Risk All Ages | DBRP | DBRP | DBRP | Y | Y | Y | DBRP | N | N | Y | DBRP | Y |

FIGURE I-9 – NOTES
CONDITIONAL MATRIX
CRITERIA FOR PULMONARY FUNCTION TEST

\*    Higher Risk – (includes one or more of the factors listed below)

- Smoker or former smoker
- History of allergies, asthma or previous occupational exposure
- Symptomatic

\*\*   Average Risk – (all factors must apply)

- Non-smoker
- No medical problems, history of medical problems, or occupational exposure
- No symptoms

Note:  If all Average Risk factors apply and one or more Higher Risk factors apply, the patient should be considered Higher Risk.

Smoker:  Current routine use of at least one cigarette/day, or one cigar/day, or one pipeful/day, or one joint (marijuana)/day.

Former Smoker:  History of at least one cigarette/day, or one cigar/day, or one pipeful/day, or one joint (marijuana)/day on a routine basis for one year or more.

PP = Pre–Placement evaluation
INT = Interval evaluation
TER = Termination evaluation

DBRP = Determined by Responsible Physician
Y   = Yes
N   = No

FIGURE I-10

CONDITIONAL MATRIX
CRITERIA FOR RESTING EKG EVALUATION

| DESCRIPTION OF EXAMINATION | EXECUTIVE EXAMINATIONS | | | ROUTINELY EXPOSED | | | NON-EXPOSED | | | SPECIAL EXPOSURE | | |
|---|---|---|---|---|---|---|---|---|---|---|---|---|
| | PP | INT | TER | PP | INT | TER | PP | INT | TER | PP | INT | TER |
| Subjective or Objective Findings, or Abnormal EKG: | | | | | | | | | | | | |
| All Ages | Y | Y | Y | Y | Y | Y | Y | DBRP | Y | Y | Y | Y |
| Normal EKG Within 6 Months and Can Provide EKG Results: | | | | | | | | | | | | |
| Age Under 30 | N | N | N | N | N | N | N | N | N | N | N | N |
| Age 30–50 | N | N | N | DBRP | N | N | N | N | N | DBRP | N | N |
| Age Over 50 | Y | DBRP | Y | Y | DBRP | Y | DBRP | DBRP | DBRP | Y | DBRP | Y |
| Never Had an EKG: No Subjective or Objective Findings; or Can't Provide Recent EKG: | | | | | | | | | | | | |
| Age Under 30 | Y | N | N | Y | DBRP | DBRP | DBRP | DBRP | DBRP | Y | DBRP | DBRP |
| Age 30–50 | Y | DBRP | DBRP | Y | DBRP | Y | DBRP | DBRP | DBRP | Y | DBRP | Y |
| Age Over 50 | Y | Y | Y | Y | DBRP | Y | DBRP | DBRP | DBRP | Y | DBRP | Y |

FIGURE I-10 – NOTES
CONDITIONAL MATRIX
CRITERIA FOR RESTING EKG EVALUATION

PP = Pre-Placement evaluation
INT = Interval evaluation
TER = Termination evaluation

DBRP = Determined by Responsible Physician
Y  = Yes
N  = No

# APPENDIX II

## THE CHALLENGE AND THE FUNDAMENTALS OF THE OCCUPATIONAL MEDICAL CAUSALITY ASSESSMENT

In an effort to assist the responsible physician and other interested parties as they confront the challenge of performing or analyzing the occupational medical causality assessment, the following elementary introduction is offered.

### Diagnostic Assessment and Identification of Causality

*Basic Concepts*

*Aggravation*: (a stimulus capable of worsening the *status quo* of a susceptible entity or condition). The concept of aggravation must be considered as either temporary or permanent (the latter is sometimes called *substantive aggravation*; the former is sometimes called *self-limited aggravation*).

*Temporary Aggravation*: the natural course (status) of an ongoing problem is temporarily worsened over time compared to what the status of the problem would have been had exposure or provocation not occurred. In cases of temporary aggravation, the patient will eventually recover to the status or condition which was predetermined by the natural history of the problem.

*Permanent Aggravation*: the natural course (status) of an ongoing problem is permanently worsened over time compared to what the status of the problem would have been had exposure or provocation not occurred. In cases of permanent aggravation, the natural course of the pre-existing condition is forever altered.

**Figure II-1** illustrates these concepts. Within the context of employment, consider the status of an individual's pre-existing problem at time $t_0$, (*pre-placement*). Following hire, if no aggravation occurs, the status of the worker's problem will follow line *A* over time - *the natural history of the pre-existing condition.* If *insult* resulting in aggravation occurs at time $t_1$, the status of the pre-existing condition will follow either line *B* (pathway of temporary or self-limited aggravation) or line *C* (pathway of permanent or substantive aggravation). If the status of the problem follows line *B*, then *recovery* will occur at time $t_2$, that point at which the status of the problem will be the same as it would have been had insult not occurred*. If the status of the problem follows line *C*, then, there will be no recovery. In theory, whether pathway *B* or *C* is followed, one can quantify the level of compensable impairment of status or function by subtracting the status on the respective curves (*B* or *C*) from that on curve *A* at any given point in time. Such an analysis, for example, may be performed routinely at time $t_3$, *termination of employment.*

*Cause:* An agent, circumstance, or event which is capable of producing a new effect or aggravating an ongoing (pre-existing) effect.

*Effect:* A diagnosis, status, function, or condition which can result from or be aggravated by a cause.

*Medically Probable:* The notion that it is *more* likely than not that something is true, from a medical standpoint.

*Medically Possible:* The notion that it is *less* likely than not that something is true from a medical standpoint.

---

* Note that status or function at time $t_2$ may be *naturally* less than it was at time $t_1$. This notion is sometimes overlooked when quantifying job-related decrements in status or function.

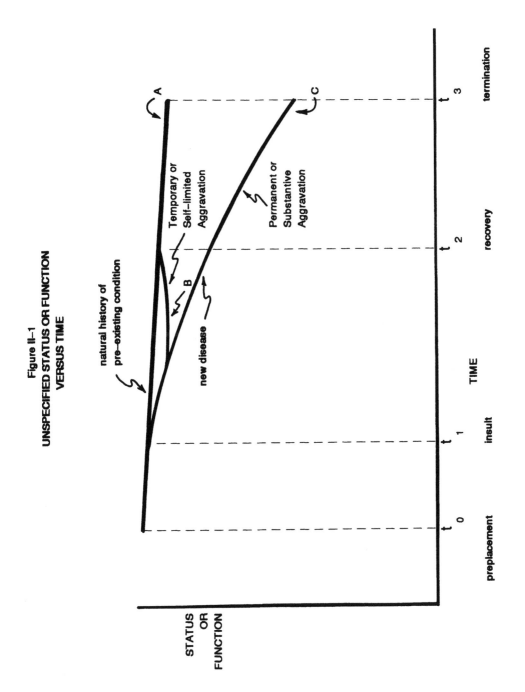

Figure II-1
UNSPECIFIED STATUS OR FUNCTION
VERSUS TIME

**Premise that a given cause (A) and a given effect (B) are associated within a reasonable degree of medical probability:**

This premise forms the core of an occupational medical causality assessment. It is the premise which the practitioner is asked to analyze or to develop and then to refute or to support in most compensation hearings. If the practitioner promotes the premise that, *within a reasonable degree of medical probability A and B are causally related*, all three of the following notions are assumed to be correct:

*A is medically probable* - A is more likely than not the cause and/or aggravator of the problem.

*B is medically probable* - B is more likely than not the correct diagnosis or condition.

*A and B are related* in a medically probable manner.

If either *A* or *B* or both is considered to be *possible* but not *probable*, then the causal association can *not* be upheld as being medically probable. Further, no number of possible causes can be taken together and viewed as a *probable* cause. Clearly, the same notion applies to *possible* effects.

Once it has been established that both *A* and *B* are *probable*, then, there must also be a *probable relationship* established between the two, before a final causality conclusion can be promulgated.

**Errors in Causality Reasoning**

A number of errors in reasoning are often committed when performing a causality analysis in occupational medicine. One of the most common involves relating one or more *possible causes* to a *probable effect* and, then, correlating and promoting these notions as forming a *medically probable cause and effect relationship*. Sometimes, a causality analysis is so deficient that a *possible cause* and a *possible effect* are correlated in a *probable* manner. This unfortunate error is the enemy of reason and should not be tolerated in occupational medicine.

In the practice of occupational medicine, it is imperative that the practitioner defer any conclusion regarding the strength of a cause and effect relationship until a satisfactory determination can be made, regarding whether the status of *medically probable* or *medically possible* better applies.

In most cases in the occupational medical setting, the diagnostic specification of problems (effects) will be very similar to the same process in the non-job-related practice of medicine. The identification of medically probable cause and effect relationships will ordinarily be a logical, straight-forward process as long as a complete examination is conducted (usually including the workplace) and a thorough, accurate history is obtained.

In less frequent, but often critical situations, however, a case will present in which the determination of probable effect (diagnosis) is simple, while the identification of probable cause is elusive. Further complications may arise even after a number of possible causes are identified, because of the difficulty of separating the non-job-related aspects of cause from the job-related ones.

*Example*

Consider an occupational medical practitioner who is faced with determining the etiology of chronic asthma in a sand and gravel truck driver who smokes two packs of cigarettes a day and who keeps three cats at home against his family doctor's advice. *Probable effect*, in this example, is the condition of chronic asthma of some type. Job-related dust exposure, smoking, and cat dander exposure may all be environmentally *possible causes or aggravators* of the asthmatic condition. On the other hand, the *true probable cause(s)* may be multifactorial, involving both environmental and intrinsic, genetically mediated factors which have not yet been identified. If the condition is one of pure intrinsic asthma, for example, the condition may not improve at all, even if the trucker changes jobs, stops smoking and gives away his cats.

It appears safe to say that within the scope of the small amount of data presented, the above example may represent one in which there are a number of *possible causes* for a *probable effect*. It would be inaccurate, without obtaining further detailed information, to conclude that any possible cause, taken alone or together with any other possible cause, can be associated with the asthmatic condition in a medically probable manner. (An obvious exception to this statement would occur, if it can be demonstrated, to a reasonable degree of medical probability, that any of the *causes* under consideration has temporarily or permanently aggravated the patient's asthma.)

## The Error of Toxicological Speculation

Clearly, the status of *possible cause* may be strengthened and promoted to the status of *probable cause*, once more detailed information is obtained. This concept may be best understood from the standpoint of toxic substance exposure. The generic presence of a toxic material, $M$, in the workplace, for example, automatically qualifies $M$ as a possible cause in many situations. $M$, however, may be biologically unavailable to the employee, or be present in quantities insufficient to cause clinically measurable effects. On the other hand, environmental or personal work-practice data may reveal that $M$ is biologically available and is present in sufficient amounts to cause or to aggravate observed problems. The error of inappropriate toxicological speculation arises when the practitioner fails to pursue the best available quantification of the biological availability of substance $M$, when $M$ is implicated as a possible cause of a probable effect.

Without sound quantifiable data, the promotion of $M$ from possible to probable cause is erroneous.

Unfortunately, in the process of seeking an explanation for an observed effect, many practitioners stumble upon possible or even remotely possible causes and, without further consideration, promote these causes to the realm of medically probable ones. This error of toxicological speculation helps no one including the patient, because it may detract from the search for true and, perhaps, preventable causes.

*Example*

Consider the illustration of an employee working during the winter in a spacious environment, heated by an allegedly, poorly ventilated gas heater, who notices severe headaches while at work. Measurement of carboxyhemoglobin reveals a level of 12%. The employee smokes two packs of cigarettes per day and drives a fifteen-year-old station wagon in poor repair. Measurement of carbon monoxide in the workplace air has not been performed and is not ordered. The employee's family physician concludes that the patient's headaches are secondary to the carbon monoxide in the work environment and the patient is sent home until the workplace is cleaned up. Two weeks later, the employee and her daughter succumb to carbon monoxide poisoning while watching a double feature drive-in movie in sub-freezing temperatures. Enough said for the pitfalls of toxicological speculation.

**Summary Imperatives**

- Remember, the notion of *probable cause* (exposure) is comprised of subset concepts, which include: the probable cause of a new problem (effect or diagnosis), the probable, temporary aggravation of an ongoing (pre-existing) problem and the probable, permanent aggravation of an ongoing (pre-existing) problem.

- Before drawing any conclusions, try to estimate the strength of your notion of cause, effect, and the relationship between the two in terms of the concepts of medical *possibility* and medical *probability*.

- Before asserting that there is a medically probable cause and effect relationship present, be certain that both cause and effect, alone, can be considered medically *probable*. Do not combine any number of possible causes together to create the overall notion of probable cause. This same imperative applies when considering possible effects or diagnoses.

- Avoid the error of toxicological speculation - namely, do not overestimate the strength of suspected cause, unless sufficient history or environmental data is present to convince you that the cause is medically probable.

- Obtain the facts regarding exposure potential - *all of the facts* - from the patient, the employer, the industrial hygienist, the laboratory, and your workplace walk-through. Without these facts, analysis of cause is useless. Surprisingly, one may spend as much or more time in this effort as one expends in formulating the patient's diagnosis.

- Finally, when in doubt, call a specialist in occupational medicine or toxicology for advice.

# APPENDIX III

## CAVEATS TO CONSIDER
## WHEN PERFORMING A WALK-THROUGH
## IN A SMALLER COMPANY

- Smaller workplaces may be less prepared or able to provide requested technical information.

- Company policies, both written or otherwise, may not exist.

- Very small workplaces (less than 10 workers) may not be legally responsible for meeting or adhering to certain regulations (e.g., OSHA). Under these conditions, the PATHMAX™ method for utilizing gathered data (see Appendix I) may, therefore, represent only a general guideline for responsible behavior, rather than a legal standard of performance.

- Smaller workplaces may not have definable unit operations or any written job descriptions. Under these circumstances, Matrices I and II in the PATHMAX™ methodology will not apply and the physician may begin by addressing Matrix III. Sometimes, when attempting to apply PATHMAX™ methodology to smaller companies, even Matrix III may be difficult to construct because job descriptions may not be written or clear and may overlap to such a degree that *risk of exposure on a specific job* may really mean *risk of exposure to all workplace hazards*.

# APPENDIX IV
## A PHYSICIAN'S WALK-THROUGH CHECK LIST

Date: _____

Client Name: _____

Contact Personnel: _____

Address: _____

_____

Telephone: ( ) _____

Fax: ( ) _____

Time Start: _____

Time Finish: _____

## PRE-WALK-THROUGH MEETING

*ITEMS DISCUSSED AND/OR REVIEWED*
(Indicate those items which are *Not Applicable* as *NA*)
(Indicate those items which are *Non-Existent* as *NE*)

1. _____ Purpose of Walk-Through:

_____
_____
_____

2. _____ Final Report: [ ] Format  [ ] Scope

[ ] Completion Date _____

[ ] Recipient  _____

_____
_____
_____

3. _____ Functional operations:  [ ] Written  [ ] Verbal  [ ] Pending

_____
_____
_____

4. _____ Unit operations:  [ ] Written  [ ] Verbal  [ ] Pending

_____
_____
_____

5. _____ Job descriptions:        [ ] Written  [ ] Verbal   [ ] Pending

_____

_____

_____

6. _____ MSDS obtained:        [ ] Yes        [ ] No

_____

_____

_____

7. _____ Health policies:        [ ] Written  [ ] Verbal   [ ] Pending

_____

_____

_____

8. _____ Safety policies:        [ ] Written  [ ] Verbal   [ ] Pending

_____

_____

_____

9. _____ Industrial hygiene
            policies:        [ ] Written  [ ] Verbal   [ ] Pending

_____

_____

_____

10. _____ Major health/safety problems (recent/historical) and audits
            (internal/external):

_____

_____

_____

11. _____ Major health/safety problems current:

_____

_____

_____

12. _____ Medical data collection/recordkeeping methods:
            [ ] Records available   [ ] Records reviewed

_____

_____

_____

13. _____ Industrial hygiene data collection/recordkeeping methods:

        [ ] Records available   [ ] Records reviewed

_____
_____
_____

14. _____ Safety data collection/recordkeeping methods:

        [ ] Records available   [ ] Records reviewed

_____
_____
_____

15. _____ Employee demographics: No. of shifts _____

        Population by shift_____

        Population by gender _____

        Population by age _____

_____
_____
_____

16. _____ Quality assurance programs and procedures:

        [ ] Programs available  [ ] Programs reviewed

_____
_____
_____

17. _____ Worker education and training programs:

        [ ] Programs available  [ ] Programs reviewed

_____
_____
_____

18. _____ Other Topics:

_____
_____
_____

## THE WALK-THROUGH

19. _____ Appropriate safety equipment available to reviewer:

       [ ] Personally provided   [ ] Employer provided

       _____
       _____
       _____

20. _____ Hazards identified relative to presence and biological availability:

       [ ] Chemical  [ ] Radiological  [ ] Biological  [ ] Physical

       _____
       _____
       _____

21. _____ Hazards correlated with:  [ ] Functional operations  [ ] Unit operations

       [ ] Job descriptions  [ ] Physical locations

       _____
       _____
       _____

22. _____ Sketch of workplace:  [ ] Completed  [ ] Provided

       _____
       _____
       _____

23. _____ Ergonomic problems identified:

       _____
       _____
       _____

24. _____ Ergonomic problems correlated with:  [ ] Functional operations

       [ ] Unit operations [ ] Job descriptions  [ ] Physical locations

       _____
       _____
       _____

25. _____ Potential psychological problems (e.g., stress) identified:

       _____
       _____
       _____

26. _____ Potential psychological problems correlated with:

       [ ] Functional operations  [ ] Unit operations

       [ ] Job descriptions  [ ] Physical locations

_____

_____

_____

*INQUIRE ABOUT AND/OR EVALUATE*:

27. _____ Temperature/Relative Humidity - **Comfort:**

_____

_____

_____

28. _____ **Facility condition and maintenance:**

_____

_____

_____

29. _____ Appropriateness of employee **attire** and **personal protection equipment:**

_____

_____

_____

30. _____ **Lighting/Harmful Electromagnetic Frequencies:**

_____

_____

_____

31. _____ **Ventilation** - general and local:

_____

_____

_____

32. _____ **Housekeeping:**

_____

_____

_____

33. _____ **Sanitary facilities:**

_____

_____

_____

34. _____ Background **noise** and **vibration**:

_____

_____

_____

35. _____ Adequacy and clarity of **signage** and **labels**:

_____

_____

_____

36. _____ Presence and condition of **chemical containers**:

_____

_____

_____

37. _____ Condition, adequacy and location of **chemical hazard storage**:

_____

_____

_____

38. _____ Location, designation and accessibility of **confined spaces**:

_____

_____

_____

39. _____ Condition, location and signage of **emergency** and **safety equipment**:

_____

_____

_____

40. _____ **First aid kits** - contents and control

_____

_____

_____

41. _____ **Lifting** and **materials handling** requirements

_____

_____

_____

42. _____ **Employee behavior** in the workplace:

_____

_____

_____

43. _____ **Self-reported problems** by medical reviewer:
_____
_____
_____

44. _____ Presence of serious or **IMMEDIATE DANGERS TO LIFE OR HEALTH:**
_____
_____
_____

45. _____ Action taken relative to # 44:
_____
_____
_____

46. _____ Other items:
_____
_____
_____

## CLOSING MEETING

47. _____ New information requested:   [ ] Date to be obtained _____
_____
_____
_____

48. _____ Pre-walk-through information: [ ] Refined  [ ] Deleted  [ ] Added
_____
_____
_____

49. _____ Future plans:  [ ] Repeat Walk-Through: Date _____
         [ ] Future work assigned: specify _____
_____
_____
_____

50. _____ Other topics:
_____
_____
_____

# APPENDIX V

## ADDITIONAL TOPICS TO BE DISCUSSED AT THE PRE-WALK-THROUGH ORIENTATION SESSION, AS EACH MAY APPLY

- Employee demographics by shift, gender, and age with consideration of temporal (e.g., seasonal) variation.

- Chronological development and performance of industrial hygiene and safety programs.

- Specification of on-site or off-site medical services including location, staffing, experience, training, specialization, and current performance.

- Discussion of overall workplace facility design including size, historical modifications, and plans for future changes.

- Discussion of the technical specifications of ventilation, lighting, and acoustical design among other systems.

- More specific recordkeeping practices, including medical records, first aid logs, medication logs, and industrial hygiene data, along with the location, format, accessibility, and management of these records.

- Quality assurance programs and procedures, with emphasis upon the fate of recommendations, internal audits, program management, and staffing.

- Worker education and training programs including design, funding, and implementation with special attention to regulatory compliance.[31]

# APPENDIX VI

## WALK-THROUGH IMPERATIVES: TECHNICAL GUIDELINES

### (VI-1) - Hazards

A hazard in the workplace is any chemical, biological, radiological, or physical agent, energy or situation which is capable of causing direct or indirect injury, primarily to an employee or an employee's offspring or secondarily to a non-employee, by whatever means (e.g., contact with the employee, employee clothing, workplace waste, emissions, etc.).

### (VI-2) - Chemical Emissions

In order for the physician to approach the challenging task of evaluating the significance of chemical emission levels, it is essential to become acquainted with some of the most common definitions used by the industrial hygienist.

*Definitions*

*TWA = Time-Weighted Average*

This term refers to the mathematical procedure of recording the concentration of a chemical substance, usually in air, (ppm, $mg/m^3$, etc.) over time and then determining the *average concentration over a specified time interval*. Mathematically, this process is equivalent to integrating the concentration versus time curve over the time interval of interest and then dividing by the magnitude of the time interval. For purposes of standardization and regulation, the TWA is usually calculated for an 8-hour time period which represents the average working day.

In general:

$$TWA = [(C_1 t_1) + (C_2 t_2) + (C_3 t_3) + ...(C_n t_n)]/[(t_1) + (t_2) + (t_3) + ...(t_n)] \qquad EQ[VI\text{-}1]$$

where $t_n$ = the $n^{th}$ time interval (n = 1, 2, 3,... etc.)

and $C_n$ = the average concentration during the $n^{th}$ time interval

*EXAMPLE*:

Assume a worker is exposed to a substance at 100 ppm for 3 hours, 120 ppm for 3 hours, and 140 ppm for 2 hours.

Then:

$C_1$ = 100 ppm and $t_1$ = 3 hours

$C_2$ = 120 ppm and $t_2$ = 3 hours

$C_3$ = 140 ppm and $t_3$ = 2 hours

Substituting into EQ[VI-1]:

TWA = [(100 x 3) + (120 x 3) + (140 x 2)]/8
TWA = 117.5 ppm

*TLV = Threshold Limit Value*

These are the registered, trademarked initials owned by the ACGIH (American Conference of Governmental Industrial Hygienists). The TLV relates to recommended exposure limits for specific chemical substances as time-weighted averages (TWA) in air. The TLV is that time-weighted average concentration to which an ordinary worker may be exposed for a normal 8-hour workday and a 40-hour workweek, repeatedly day after day without adverse effect.[32,33] The TLV should *not* be interpreted as a legal regulation, nor as an absolute cut-off between hazardous and non-hazardous levels of exposure. Workers who are more susceptible or even hypersensitive to selected agents may respond adversely to levels below the TLV, and more hardy workers may experience no problems until exposure levels exceed the TLV. The TLV should not be confused with the PEL or the REL, both of which, although TWAs, are terms utilized by OSHA and NIOSH, respectively. TLVs, PELs, and RELs may all be quantitatively different in value.

A useful formula to keep handy allows you to convert the TLV in *ppm* to *mg/m³* in air at 25°C and one atmosphere.  It is:

TLV in mg/m³ = [(TLV in ppm) (GM$_r$)/24.45]                    EQ [VI-2] [34]

where:

GM$_r$ = gram molecular weight of the substance

The term TWA may be used with another abbreviation to specify the method by which other limits are calculated.  For example, the term TLV may be written as TLV-TWA, which means that the Threshold Limit Value is considered, conceptually, to be a Time-Weighted Average.  If, for example, the TLV for the solvent toluene is 100 ppm, then the average concentration of toluene over a normal 8-hour workday and a 40-hour workweek should not exceed 100 ppm, as a Time-Weighted Average.

Both practically and theoretically speaking, it should be apparent that the TLV-TWA is only one of several possible exposure limit parameters, which could be considered, when specifying acceptable exposure conditions.  For example, consider the mathematical fact that it is possible for the airborne concentration of a chemical to rise from 3 to 5 times above the TLV-TWA level for a few seconds or even a few minutes without causing the calculated TWA level to rise above the TLV-TWA.

*EXAMPLE:*

Assume the TLV-TWA is 100 ppm and a worker is exposed to 80 ppm for 4 hours, 500 ppm for 0.1 hours, 100 ppm for 0.5 hours, and 60 ppm for 3.4 hours.

Substituting into EQ[VI-1]:

TWA = [(80 x 4) + (500 x 0.1) + (100 x 0.5) + (60 x 3.4)]/8

TWA = 78.0 ppm

In this example, the exposure level was five times the TLV-TWA for 6 minutes, but the calculated TWA for the workday fell well below the

TLV-TWA. Because of the situation illustrated in the above example, it should be clear that when establishing acceptable exposure conditions, corollary exposure limit parameters must also be specified which define just how high an exposure concentration can go (if even for a short time) without causing problems for the worker. There are three basic parameters utilized by the ACGIH and acknowledged, in principle or in practice, by OSHA and NIOSH for this purpose. They are the:

*STEL = Short-Term Exposure Limit*

> The concentration to which workers can be exposed continuously for a short period of time without suffering from 1) irritation, 2) chronic or irreversible tissue damage, or 3) narcosis of sufficient degree to increase the likelihood of accidental injury, impair self-rescue or materially reduce work efficiency, and provided that the daily TLV-TWA is not exceeded. It is not a separate independent exposure limit; rather, it supplements the time-weighted average (TWA) limit where there are recognized acute effects from a substance whose toxic effects are primarily of a chronic nature. STELs are recommended only where toxic effects have been reported from high short-term exposures in either humans or animals.

> A STEL is defined as a 15-minute TWA exposure which should not be exceeded at any time during a workday even if the 8-hour TWA is within the TLV-TWA. Exposures above the TLV-TWA up to the STEL should not be longer than 15 minutes and should not occur more than four times per day. There should be at least 60 minutes between successive exposures in this range. An averaging period other than 15 minutes may be recommended when this is warranted by observed biological effects.[32]

*TLV-C = Threshold Limit Value - Ceiling*

> ... the concentration that should not be exceeded during any part of the working exposure.

> In conventional industrial hygiene practice, if instantaneous monitoring is not feasible, then the TLV-C can be assessed by sampling over a 15-minute period except for those substances that may cause immediate irritation when exposures are short.

> For some substances, e.g., irritant gases, only one category, the TLV-Ceiling, may be relevant. For other substances, one or two categories may be relevant, depending upon their physiologic action. It is important to observe that if any one of these types of TLVs is exceeded, a potential hazard from that substance is presumed to exist.[32]

*Excursion Limits*

The excursion limit provides guidelines on just how high an exposure concentration can go for those substances which have neither a STEL nor a TLV-C limit.

According to the ACGIH:

> For the vast majority of substances with a TLV-TWA, there is not enough toxicological data available to warrant a STEL. Nevertheless, excursions above the TLV-TWA should be controlled even where the 8-hour TLV-TWA is within recommended limits.[35]

For those substances with TLV-TWAs for which no STEL or Ceiling is specified:

> Excursions in worker exposure levels may exceed 3 times the TLV-TWA for no more than a total of 30 minutes during a workday, and under no circumstances

should they exceed 5 times the TLV-TWA, provided that the TLV-TWA is not exceeded.[35]

If a STEL is established for a chemical, the published value takes precedence over the excursion limit recommendation.

## Skin Notations

Many substances in the TLV list are given a *skin* notation. This notation means that the ACGIH considers direct or airborne exposure to the skin, mucous membranes, or the eye as significant routes of entry and potential toxicity to the body. The ACGIH states:

> Little quantitative data are available describing absorption of vapors and gases through the skin. The rate of absorption is a function of the concentration to which the skin is exposed.

> Substances having a skin notation and a low TLV may present a problem at high airborne concentrations, particularly if a significant area of the skin is exposed for a long period of time. Protection of the respiratory tract, while the rest of the body surface is exposed to a high concentration, may present such a situation.

> Biological monitoring should be considered to determine the relative contribution of dermal exposure to the total dose.

> This attention-calling designation is intended to suggest appropriate measures for the prevention of cutaneous absorption so that the TLV is not invalidated.[36]

## PEL

*Permissible Exposure Limit* is a TWA "that must not be exceeded during any 8-hour work shift of a 40-hour work-week."[37] The STEL is intended to complement the PEL.[38]  The PEL is the abbreviation used by OSHA to designate regulated TWA values for specific hazards.

## REL

*Recommended Exposure Limit* is a TWA promulgated by NIOSH and refers to limits which should not be exceeded over an 8 or 10-hour day (individual RELs will specify which) for a 40-hour work week.  The REL designation can also be applied to TWA ceiling levels with exposure time limits ranging from instantaneous to 120 minutes.[39]

## IDLH

*Immediately Dangerous to Life or Health* level is a level utilized by OSHA and NIOSH that represents:

> The maximum concentration from which, in the event of respirator failure, one could  escape within 30 minutes without a respirator and without experiencing any escape-impairing (e.g., severe eye irritation) or irreversible health effects.[40]

## AL

*Action Level* is the TWA exposure level below the PEL or REL but above which medical surveillance should commence.   If true, representative exposure levels are consistently below the Action Level, medical surveillance may be optional.  The Action Level is often set at 50% of the PEL or REL.

## BEI

*Biological Exposure Index* is another term used by the ACGIH and refers to biological monitoring of workers or, namely, that monitoring in which specimens of blood, serum, urine, saliva, fat, stool, expired air, or other biological materials may be obtained for analysis.  A suspected chemical, metabolite or some other biological consequence of exposure may be measured during biological monitoring.   The ACGIH states:

> ... BEIs represent the levels of determinants which are most likely to be observed in specimens collected from a healthy worker who has been exposed to chemicals to the same extent as a worker with inhalation exposure to the TLV.   BEIs do not indicate a sharp distinction between hazardous and nonhazardous exposures.   Due to biological variability it is possible for an individual's measurements to exceed the BEI without incurring an increased health risk.   If, however, measurements in specimens obtained from a worker on different occasions persistently exceed the BEI, or if the majority of measurements in specimens obtained from a group of workers at the same workplace exceed the BEI, the cause of the excessive values must be investigated and proper action taken to reduce the exposure.
>
> BEIs apply to eight-hour exposures, five days a week. However, BEIs for altered working schedules can be extrapolated on pharmacokinetic and pharmacodynamic bases.   BEIs should not be applied either directly or through a conversion factor, in the determination of safe levels for nonoccupational exposure to air and water pollutants, or food contaminants.   The BEIs are not intended for use as a measure of adverse effects or for diagnosis of occupational illness.[41]

Some good examples of readily understandable BEIs would be those for carbon monoxide (CO):   Namely, carboxyhemoglobin in blood should be less than 8% and CO in end-exhaled air should be less than 40 ppm.   As another example, consider fluorides in urine:   prior to shift, the BEI should be less than 3 mg/L and at the end of shift the BEI should be less than 10 mg/L.[42]

*Chemical Mixtures*

It should be emphasized that chemical emissions may be comprised of single compounds or mixtures. When evaluating the significance of these hazards, remember that even though exposure limits may not be exceeded for individual components of a mixture, the exposure limit for the mixture as a whole may be unacceptable if, collectively, a number of individual component exposure levels are relatively high. An industrial hygienist can assist in calculating the acceptable exposure level for mixtures. For those mathematically inclined, however, here is the methodology:

The TLV for certain chemical mixtures is exceeded if the sum of the fractions (ratios) of each individual chemical component concentration (C) divided by its respective TLV exceeds unity. This relationship holds, if each chemical adversely affects the same organ system or works with the others in some additive fashion. For example, if a mixture contains three chemical components (1, 2, and 3) each present at average concentrations $C_1$, $C_2$, and $C_3$, over a specified time interval and if:

$$(C_1/TLV_1) + (C_2/TLV_2) + (C_3/TLV_3) > 1 \qquad \text{EQ[VI-3]}$$

where:

$TLV_1$, $TLV_2$ and $TLV_3$ are the respective TLVs of components 1, 2, and 3,

then:

the TLV of the mixture is exceeded.[43] Again, this formula applies when the individual components are thought to have additive, *not* synergistic, effects. In the latter case, the TLV of the mixture may be exceeded when the sum of ratios is less than 1.0.

Assuming *additive* effects, consider the following:

*EXAMPLE*:

If:

$C_1 = 50$ ppm ; and $TLV_1 = 200$ ppm
$C_2 = 60$ ppm ; and $TLV_2 = 300$ ppm
and
$C_3 = 10$ ppm ; and $TLV_3 = 20$ ppm
then substituting into the left side of EQ[VI-3]:

$$(C_1/TLV_1) + (C_2/TLV_2) + (C_3/TLV_3) = 50/200 + 60/300 + 10/20$$
$$= 0.25 + 0.20 + 0.5$$
$$= 0.95$$

Since 0.95 is less than 1.0 (see EQ[VI-3]), the TLV of this mixture is not exceeded.  Note, that if the concentration of $C_2$ had been only about 80 ppm, the TLV of the mixture would have been exceeded, even though no single component exceeded 50% of its own TLV.

If components of a mixture act completely *independently* of each other (e.g., adversely affect entirely different organ systems), then the TLV of the mixture will be exceeded only if the concentration of a single component exceeds its own TLV.[44]*

It is important to re-emphasize that some chemical mixtures may contain individual components which can potentiate the toxicological effects of one another (for example, carbon black or nuisance dust particulates and sulfur dioxide).  One must exercise care, therefore, when comparing TLVs or PELs with measured exposure levels of such mixtures.

*Particulate Exposures*

When evaluating the significance of particulate exposures, among other things, it is important to differentiate between the respirable

---

* The obvious pitfall to this logic is that *independent* adverse effects on different organ systems, may have, in reality, additive or even synergistic effects physiologically.

and non-respirable fractions of the aerosol. The toxicological effects upon the worker from each fractional component may differ greatly.

Generally speaking,[45] the respirable fraction of particulates includes those with median diameters of less than 10 microns (1 micron = $10^{-6}$ meters). Particle deposition in the airways is a very complex statistical, anatomical, chemical, and physical process. For this reason, some particles greater than 10 microns (especially long fibers with small cross sections) will still manage to become trapped in the lungs, although most will be trapped in the nasopharynx. The overall alveolar percentage deposition of particles less than 10 microns in size tends to increase with decreasing size until particle size drops below about 3-4 microns. Below this size range, overall alveolar percentage deposition decreases (less retention) with decreasing size until particle sizes reach about 0.2 to 0.3 microns. Below this size range, percentage deposition begins to rise slightly according to some mathematical models,[46] although according to other models the alveolar deposition fraction remains fairly constant (about 20%) down to sizes of about 0.1 micron.

*The Concept of Dose as a Function of Exposure Level and Exposure Time*

As is the case for all hazards, it should be emphasized that the exposure level of a chemical mixture or individual chemical substance is not the sole consideration in evaluating the toxicological risk to the worker. Remember that the actual dose (D) to the worker (assuming intake via inhalation, ingestion, or skin absorption) is directly proportional to both the exposure level concentration *and* the duration of exposure.

*EXAMPLE*:

For inhalation:

$$D = C \times V_R \times K \times t \qquad \text{EQ[VI-4]}$$

where:

D = absorbed dose in mg

C = average exposure level concentration in mg/m³
        over the time duration of exposure

$V_R$ = pulmonary ventilation rate in m³/min

K = fraction absorbed

t = exposure time in minutes

Now, assume exposure to lead fumes, where:

C = 0.05 mg/m³ (TLV-TWA = 0.15 mg/m³)

$V_R$ = 0.016m³/min or (16 1/min)

K = 0.2

t = 300 min

then substituting into EQ[VI-4]:

D = (0.05 mg/m³) x (0.016 m³/min.) x (0.20) x (300 min.)

D = 0.048 mg = 48 micrograms (the actual amount of lead
        absorbed by the worker, via inhalation after 300 minutes
        of exposure)

Reasoning from U.S. Environmental Protection Agency recommended Maximum Concentration Levels (MCL) for lead in drinking water (50 micrograms/liter)[47] and assuming an average consumption of 2 liters of water/day, the average daily intake of lead should not exceed about 100 micrograms. As indicated, in the hypothetical scenario described above and excluding intake from other than airborne sources, the exposed worker would absorb only 48 micrograms of lead during 300 minutes or 5 hours of exposure time. If exposure time were doubled to 600 minutes (10 hours) and all other factors remained unchanged, however, the worker would absorb 96 micrograms, even though the airborne exposure level was only 1/3 of the TLV-TWA. If we now consider even a small amount of additional lead absorption from drinking water, it is clear to see that total lead absorption could easily exceed the 100 microgram limit recommended by the EPA.

Using this simple example, one can see that exposure to supposedly acceptable airborne concentrations of a hazardous material for too long a time may lead to an unacceptable absorbed dose to the worker, when all factors are taken into account. By the same token, it should be emphasized that brief exposures to relatively high concentrations of some chemicals may impart an acceptable, low dose exposure to the worker (as long as instantaneous concentrations do not exceed ceiling limits or other recommended maxima).

The example outlined above illustrates one of the reasons why TLVs, PELs, STELs, and RELs are all exposure concentrations to which an *acceptable exposure time limit* is assigned.

In summary, when performing the walk-through and assessing the significance of chemical exposure levels, do not forget to estimate the potential total dose (not just the concentration) of the hazard to the worker, based upon the route of exposure, the efficiency of absorption, and the time duration of the insult.

## (VI-3) - Skin Exposure

Many hazardous materials can be absorbed through the skin especially if the skin is injured or abraded. This phenomenon is especially true for solvents which may damage the skin by defatting the epidermal layers (See VI-2: *Skin* notation).

## (VI-4) - Ergonomics

*Ergonomic* (from *ergon*, meaning *work* in Greek) is an adjective derived from the noun *ergonomics*, which is the scientific discipline dedicated to the study of the interrelationships between human workers and their physical work environment. Ergonomics is a multidisciplinary field which addresses not only physical but also procedural, methodological, and psychological factors. An ergonomic problem is a problem which may be caused by a mismatching of the individual with respect to any aspect of one or more of these factors and which may be mitigated by procedural or physical means.

There has been a veritable explosion of new textbooks on this subject over the past few years.  It is advisable to rely upon one or two of the classics,[48-51] when first approaching this fascinating and challenging field of study.

### (VI-5) - Psychological Stress

The physician should realize that *stress*, per se, is intrinsic to all endeavors in life, including those activities performed at work.  What differentiates *life stress* from potential worker compensation *job stress* is simply what is known as *proximate cause*.  From the physician's standpoint, with respect to stress, the concept of proximate cause relates to those situations in which both stress and the consequences of stress evolve from adverse circumstances which the worker would probably not encounter, were it not for his or her job.  For example, a medically documentable situational anxiety or panic disorder may be considered *job-related* if it occurs in a worker whose assembly-line production quotas were arbitrarily doubled over the six months prior to diagnosis and whose medical and non-work-related social and psychiatric history were documented to be negative.  An identical clinical disorder, however, which is allegedly related to a personality conflict between a worker and his or her supervisor or co-worker may not qualify as an etiologic factor in a *job-related stress claim*, because such adverse interactions may equally likely or even more likely afflict the worker in a non-job-related setting.

The evaluation of psychological stress factors must be pursued by the physician with great care and caution.  The physician should learn the local, statewide definitions of job stress and occupational disease as well as the basic worker compensation law surrounding these terms, before drawing specific conclusions about diagnosis and cause and effect relationships.

### (VI-6) - Comfort

There are three basic methods of measuring temperature in the workplace:[52]

*Dry bulb temperature* (DBT) is that obtained by an ordinary mercury thermometer.

*Wet bulb temperature* (WBT) is that obtained by placing a water wetted wick over the mercury reservoir of a dry bulb thermometer. Normally, WBT is less than DBT because of the cooling effect of evaporation.

*Globe temperature* (GT) is that obtained when the mercury reservoir of an ordinary dry bulb is placed inside a metal sphere whose exterior is painted black and through whose surface the thermometer's body or stem protrudes and is secured by a rubber stopper. This thermometer is utilized to measure radiant heat from the surroundings. GT may be increased, for example, outdoors on an asphalt surface or even on snow.

The most useful *heat stress index* is called the *wet bulb globe temperature* (WBGT) which combines the effects of dry temperature, evaporative cooling, and radiant heat.

When there is no solar load (heat from direct sun), whether indoors or outdoors:

$$WBGT = (0.7)(WBT) + (0.3)(GT) \qquad\qquad EQ[VI\text{-}5]$$

When outdoors with a solar load:

$$WBGT = (0.7)(WBT) + (0.2)(GT) + (0.1)(DBT) \qquad EQ[VI\text{-}6]$$

WBGT is the *heat stress index* which is utilized in NIOSH,[53] OSHA, and ACGIH standards and recommendations. WBGT should not be used to measure heat stress in individuals wearing impermeable clothing (e.g., fully suited hazardous waste workers) because there is little or no benefit of cooling by evaporation.

Finally, *relative humidity* (RH), which is proportional to the amount of water vapor in the air, is measured using a psychrometer or hygrometer which provides both WBT and DBT. By applying the recorded values of WBT and DBT to a psychrometric chart, the relative humidity can be graphically identified.[54]

### (VI-7) - Facility Condition and Maintenance

Any physician who intends to perform a workplace walk-through and wishes access to nicely cross-referenced and well condensed OSHA regulations should purchase the annual two-volume set called *Best's Safety Directory* (A.M. Best Company, Inc., Ambest Road, Oldwick, New Jersey 08858; (908)439-2200).

This reference is an excellent one against which you can compare many of your basic observations from the walk-through. For example, if you want to check on the subject of ladders, *Best's Safety Directory* (BSD) indicates that:

> The adequacy of ladders and the work practices
> followed by employees using them are regulated by
> OSHA in three sections:  Portable Wood ([29 CFR]
> 1910.25), Portable Metal ([29 CFR] 1910.26), and Fixed
> Ladders ([29 CFR] 1910.27).[55]

Requirements from the OSHA Construction Standards (29 CFR 1926.451) and even the Maritime Standards are also discussed[56] in detail, along with the three specific OSHA ladder standards over several pages.

The information provided by BSD should be used only to provide a frame of reference for discussions about risk management and prevention.  It is *not* recommended that the physician attempt to become an expert in OSHA standards or that the physician report findings as necessarily being within or in discordance with OSHA regulations.  The medical walk-through is *not* intended to be an OSHA audit exercise.

If you have serious questions about OSHA compliance, refer them to a qualified industrial hygienist, safety engineer, or risk manager. Leave absolute statements about OSHA compliance out of your final written or verbal report unless you are willing to assume the liability of misstatement.

## (VI-8) - Clothing and Personal Protection

Proper employee attire or clothing is a very important factor in maintaining optimum occupational health and safety. During the walk-through, the physician should pay particular attention to the following factors:

- whether employees are wearing their own clothing or company-provided uniforms.

- if personal clothing is worn, whether there is a risk of transporting workplace contaminants home.

- whether clothing presents a hazard to the employee (e.g., a drill press or lathe operator whose loose fitting sleeves can become caught in rotating machinery).

- whether improper clothing can create an unacceptable heat or cold stress to the employee.

- whether clothing has resulted in any skin irritation or other discomfort problems.

- whether clothing is adequate to protect the worker from hazardous skin exposure or whether it should be augmented with other protective barriers such as impermeable aprons or sleeves.

As a corollary to the above factors, it is important to emphasize that permeable, chemically contaminated clothing can enhance systemic absorption of many hazardous chemical agents and can foster the development of epidermal injury and contact dermatitis.

If the employee's clothing is company-provided and commercially laundered, the physician should keep in mind that some commercial industrial laundries use extremely harsh soaps and do not rinse clothing thoroughly. This phenomenon can be particularly important to the physician, who is trying to explain the etiology of a dermatitis problem, thought by the employee or others to be related to workplace chemical exposure.

Options for a wide variety of protective clothing, suitable for different exposure conditions, can be found along with a relevant OSHA regulatory summary in *Best's Safety Directory*.[57]

### Personal Protection Equipment

Be particularly observant about employee compliance regarding utilization of specific personal protection equipment in areas where such equipment is supposed to be used.  Usually such areas are designated by signs specifying, for example, hearing protection, hard hats, respirators, or eye protection.  If signs are not readily apparent and it appears that the area should be designated in some manner, based upon workplace activities or unit operations, ask your escort to clarify the situation for you.

### Gloves

You may see workers wearing aprons or gloves, usually with some form of eye protection.  It is useful to ask about the type of glove material being employed and whether that material is impermeable to the chemical agents being handled.  Often, the employer will mismatch the glove type and chemical hazard, inadvertently.  There are a number of excellent references which may help you to become better able to converse intelligently on this subject.[58,59] *Best's Safety Directory* also has an excellent section on this subject.[60]  In addition, most glove manufacturers can provide you with tables which outline the permeabilities of various glove material options, relative to the compound being handled.

Finally, remember that gloves are useless if hazardous materials work their way down inside them. When this problem occurs, the glove serves as an occlusive barrier and enhances the absorption and destructive action of the chemical. Gloves, therefore, must be long enough and sealed well enough to protect the worker. Skin allergies to glove materials or glove liners, moreover, are common, so it is necessary to remember this fact when formulating a differential diagnosis for hand or arm dermatitis.

## *Goggles or Safety Glasses*

Goggles or safety glasses with side shields are critical items in those areas in which the eyes are vulnerable to acid mists, mechanically generated particulates (e.g., from grinding wheels, chisels, sanders, etc.) and other irritating airborne chemical hazards.

Fully sealed, chemically resistant goggles must be worn by all individuals, (especially those who wear contact lenses) who are at risk for splash or chemical vapor hazards. Some companies prohibit the wearing of contact lenses in such situations, (a good policy), but many qualified workers cannot wear glasses and need to be accommodated in this regard.

## *Shoes*

Shoes with steel toes are often required in areas in which there is a significant risk of crush injury to the foot. Observe whether workers wear ordinary tennis shoes or other inadequate footwear under these conditions. Remember that not all safety shoes are equivalent in protecting the worker from injury.[61,62]

## *Hard Hats*

Hard hats must fit properly and are usually positioned so that the brim is horizontal or tilted slightly forward. Improper fit or positioning of the hard hat may render it ineffective and cause it to slip off the worker's head.

Hard hats are available with a wide assortment of specifications, related to voltage, impact, and even eye protection.[63] It is advisable to inquire whether the hard hats used on the worksite are appropriate for the demands of the job.

## Hearing Protection

Workers may wear a variety of different types or brands of hearing protection, but categorically this protective device is available as either plugs or muffs. Different manufacturers of ear plugs and muffs make claims about the attenuation capabilities of specific products as a function of noise frequency. Properly fitting muffs can attenuate about 30-40 dB at critical frequencies, namely, those in the speech range (500-3000 Hz).[64,65] Newly designed ear plugs can attenuate about 25-35dB over the same frequencies.[66] Generally speaking, therefore, muffs are better than plugs. Plugs should be disposable. If not kept clean, ear plugs can promote the development of otitis externa or ear canal infections. Muffs, on the other hand, can be reused and can be cleaned on a periodic basis. Finally, be sure to observe whether workers put cotton or other undesirable items in their ears.

## Respirators

There are many different types of respirators that are used in the workplace depending upon the job at hand. Basically, they can be classified by the following factors:

- Coverage area (e.g., half face, full face, hood).

- Positive or negative pressure. If *positive* pressure, the work of breathing by the user is assisted by air pressure; if *negative* pressure, the work of breathing is not assisted and the worker must provide the pressure differential necessary to move air through the filter.

- External (e.g., tank or compressor) air supplied (EAS) or ambient air dependent (AAD). External air supplied respirators are connected to pressurized air tanks or to compressors which move clean, fresh air from a remote location to the user. Ambient air dependent respirators provide air to the user from the immediate surroundings.

EAS respirators may operate under positive or negative pressure. If they operate under positive pressure, the flow to the respirator may be a continuous one or the air may flow only long enough to maintain a specified (valve regulated) positive pressure. If the EAS respirator operates under negative pressure, it is called a *demand airline device*.[67] This device provides breathing air only during inhalation, when a negative pressure is generated inside the respirator cavity.

All AAD respirators filter contaminated air through absorbent cartridges, sometimes with pre-filters, which are selected specifically for the hazardous materials to which exposure is anticipated.

Obviously, there are many different types of respirator combinations possible, depending upon the different characteristics stated above. The most common ones which you will see in the workplace are:

- The *negative pressure*, AAD, half or full face mask respirator (also known as a negative pressure air purifying respirator).

  *Powered by*: the breathing pressure differential of the user. The negative pressure of breathing moves air through the filter elements, into the airways.

  *Utilized for*: routine type work, for example, near a solvent degreaser or in dusty work areas involving sanding, grinding, polishing, or chemical handling.

- The *positive pressure*, AAD respirator, which may be configured with a half or full face piece hood, or helmet.

  *Powered by*: a blower which usually sucks air through the filter elements and pushes it into the airways.

*Utilized for*: the same type of work as the negative pressure air purifying respirator, when less demand on breathing is desired or respirator fit is a problem.

- The EAS respirator, or airline respirator. The continuous flow positive pressure type is quite common. It may be equipped with a half or full face mask but is usually seen with a hood or helmet.

*Powered by*: bottled, compressed air or a compressor, which moves clean air, under positive pressure into the airways.

*Utilized for*: sand or shot blasting, confined space, or tank entry, high pressure spray painting and in areas where oxygen deficiency could be a problem.

- The SCBA or self-contained breathing apparatus. This is an EAS respirator in which the compressed air tanks are carried on the back of the user. It may operate in continuous, positive pressure or negative pressure, demand mode and is usually equipped with a full face mask.

*Powered by*: bottled compressed air.

*Utilized for*: rescue, hazardous materials spill clean-up, tank or confined space entry. The SCBA is the familiar respirator worn by firefighters.

The physician should observe and evaluate respirator utilization, maintenance, and storage. Remember to inquire whether employees who have been assigned respirators have been medically certified to wear them (29 CFR 1910.134).[68] Also inquire whether the respirators have been properly fit tested for any leaks in the facial seal. In this regard, remember that any facial hair, especially beards, which may interfere with respirator facial seal, should be prohibited.

Finally, you should ask your escort if the company has carefully researched whether the chemical cartridges chosen for AAD respirators are properly matched for the airborne contaminant from which protection is sought and whether the cartridges are replaced on a regular basis to prevent *break-through* of chemical contamination into the worker's breathing zone.

## (VI-9) - Lighting

*Wavelength and Frequency*

The frequency and wavelength of electromagnetic (EM) radiation; are related as follows:

$$c = f\lambda \qquad\qquad EQ[VI-7]$$

where:

   c = the speed of light  $3 \times 10^{10}$ cm/sec in a vacuum

   f = frequency in Hertz or cycles/second

   $\lambda$ = wavelength in cm

*The Basics, including Illuminance, Luminance and Reflectance*

Workplace lighting should be safe, functional and not produce excessive eyestrain, glare, or shadows. With respect to eyestrain, for example, where fluorescent lights are widely used, you may see that some employers have chosen fluorescent bulbs which emit longer, *softer* wavelengths.  These bulbs have a more reddish or pink appearance and are represented as being more compatible with incandescent wavelengths. Some workers, however, do not like them, because they are perceived as not being bright enough.

The physician should observe both local (e.g., lighting of the work area of a drill press) and general lighting.  All lights should be protected from accidental breakage and should be placed in areas where they will not create an explosion or fire hazard.

The TLVs established for visible light include electromagnetic frequencies in the near infrared region.  The ACGIH specifies wavelengths between 4000 and 14000 Angstroms (1 Angstrom = $10^{-8}$ cm) as those for which the TLVs apply. Actual TLVs are designed to protect the retina from thermal injury and photochemical injury from chronic blue-light exposures.[69]

Visible light intensity (luminous intensity) is the light emitted by a point source, and is measured in a unit called the *candela* *(cd)*, which is defined in terms of the emission of a black body radiation source at the freezing (solidification) point of platinum, 2040 °K.[70]

The *flow rate* of light from a source is called the *luminous flux* and is measured in a unit called the *lumen*, which is defined as the luminous flux emitted from a point source of one candela intensity through a solid angle of one steradian [Remember the radian? It is the central angle of a circle which subtends a length of arc equal to the radius along the circumference of the circle. One radian equals approximately 57.3 degrees. Similar in concept, a steradian is that solid angle originating from a point at the center of a sphere of radius, r, which subtends a surface area on the sphere equal to $r^2$. Since the entire surface area of the sphere is $4 \pi r^2$, there are $4 \pi$ steradians, in total, originating from the center, as vertex ($\pi \approx 3.14159$). If a one candela source emits one lumen through each steradian, then the overall luminous flux from a point source of one candela in all directions is $4 \pi$ or about 12.57 lumens].

As noted, a point source of one candela will emit luminous flux equal to one lumen through each steradian in space. By simple solid geometry, it can be seen that the steradian will subtend a surface of one square foot in area on a sphere whose surface is one foot from the one candela light source (r = 1ft.). The *illumination* of that surface is defined to be *one foot-candle*. In other words a *foot-candle (ft-c)* is a unit of illumination equal to one lumen/ft². In the same geometric sense, the steradian will subtend a surface of one square meter in area on a sphere whose surface is one meter from the one candela light source (r = 1m). The illumination of that surface is defined to be one

---

* The unit of the *candela* is the unit of luminous intensity equal to one *candle* or one *candle-power*. The *candela* is the SI (Standard International) unit of luminous intensity.[71]

lux(l).  In other words a *lux* is a metric unit of illuminance equal to one lumen/m$^2$.[72]  Clearly in terms of illumination (and geometry), the lux must be less than the foot-candle.  In fact one lux ≃ 0.093 foot-candles.

Consider the following example: If the industrial hygienist says, "we measured an illuminance of 50 lux (or about 4.65 ft-c) on that tabletop over there," it means that the table is being illuminated by 50 lumen for every square meter (or about 4.65 lumen for every square foot) of tabletop surface area.

Remember the inverse square law, when evaluating the adequacy of lighting.  If the distance between the illuminated surface and the light source is doubled, the illumination will drop by a factor of four.  If the distance is tripled, the illumination will drop by a factor of nine.  If the distance is halved, the illumination will *increase* by a factor of four and so on.

*Luminance* (as opposed to *illuminance*) characterizes how much light is *emitted* from a surface[73] and can be thought of as the *brightness* of the surface.  The luminance of a perfectly diffuse, perfectly reflecting surface illuminated by one foot-candle is defined as a *foot-lambert (ft-l)*.  The metric unit of luminance is the *candela/m$^2$* which is also called the *nit*.

In summary, remember that *illuminance* (measured in *lux* or *foot-candles*), refers to the *extent to which the surface is illuminated*, and that *luminance* (measured in *candela/m$^2$* or *foot-lamberts*) refers to the *light coming from a surface*.

As a frame of reference, consider that the lighting (illuminance) requirements for performing various tasks in the workplace can vary by factors of several hundred.  For example, a *public space with dark surroundings* may only require 50 lux, the *performance of visual tasks of high contrast or large size*, may require 500 lux, and the *performance of very prolonged and exacting visual tasks* may require 10,000 lux of illuminance.[74]

The concept of *reflectance* refers to a comparison between the illuminance and *subsequent* luminance of a surface upon which light falls.  For a perfectly diffuse, perfect reflector, there will be no preferential direction for reflected light and no light will be lost.  If the *illuminance* of a perfectly diffuse, perfect reflector is one lux (one lumen/m²), for purposes of reflectance considerations, the *luminance* of the surface is defined as one *apostilb* (asb).

If the surface does not reflect all light, but only a fraction, X, (its reflectance), then the luminance (L) in *asb* will be equal to X times the illuminance (I) in *lux*.

Namely:

$$L = (X)(I) \hspace{4cm} EQ[VI\text{-}8]$$

*EXAMPLE*:

If reflectance, X = 0.5 and a surface has an illuminance I = 100 lux, then, substituting into EQ[VI-8]:

$$L = (0.5)(100) \text{ asb}$$

$$L = 50 \text{ asb}$$

The unit relationship equivalence between the standard units of luminance (candela/m²) and apostilbs for a perfectly diffuse, perfectly reflecting surface is: *one candela/m² is equivalent to π (asb)* **or** *there are (1/π) candela/m² per asb*.[75]

Stated as an equation:

If $C_m$ = the number of candela/m² and Asb = the number of apostilbs, then:

$$(\pi)(C_m) = \text{Asb.} \hspace{3cm} EQ[VI\text{-}9]$$

This equation is the key to deriving the relationship between the *illuminance* of a surface in lux (lumen/m²) and the *luminance* of the surface in candela/m².

If, by definition, for a perfectly diffuse, perfectly reflecting surface, one *lux of illuminance* creates one apostilb of luminance and each *apostilb of luminance* is equivalent to $(1/\pi)$ candela/m² of luminance, then this simple relationship can be expressed as:

$$L = (I)/(\pi) \hspace{5cm} \text{EQ[VI-10]}$$

where:

L = luminance in candela/m²

I = illuminance in lux (lumen/m²).

Since most surfaces are not perfectly diffuse, perfectly reflecting surfaces, the luminance of an illuminated surface may vary from the relationship just described. The equation used to express this notion is:

$$L = (I)(LF)/(\pi) \hspace{4cm} \text{EQ[VI-11]}[75]$$

where the LF is a Luminance Factor and is specified by Howarth as:

> ...the ratio of the luminance of a reflecting surface, viewed in a given direction, to that of a perfect white diffusing surface identically illuminated.[75]

Howarth continues:

> If the reflecting surface is itself a perfect diffuser, then the value of the luminance factor is the same as the reflectance, is independent of the viewing position and cannot be greater than one....if the surface does have specular reflections, then the luminance factor will vary with viewing position and at the angle of reflection could be greater than one.[75]

*EXAMPLE*:

For the case presented earlier, in which the reflecting surface is a perfectly diffuse but not a perfectly reflecting surface with a reflectance of 0.5, then LF also equals 0.5.

If:

I = 100 lux

then substituting into EQ[VI-11]:

L = (100)(0.5)/3.14

or

L = 50/3.14 = 15.91 candela/m²

*Contrast*

The *Contrast* ($C_t$) between two objects is a unitless term defined as:

$$C_t = [(L_{brighter}) - (L_{darker})]/(L_{brighter})$$   EQ[VI-12][76]

where:

$L_{brighter}$ = the luminance of the brighter object

$L_{darker}$ = the luminance of the darker object

Clearly, if ($L_{brighter}$) = ($L_{darker}$), then the contrast between the objects is zero.

Finally, the *luminance ratio* (LR) is a corollary concept to *contrast* and is simply defined as the ratio between the luminance of a task and the luminance of the background or the surface upon which the task is being performed.[77]

*Video Display Terminals (VDTs)*

VDTs are also known as VDUs (Video Display Units) or even CRTs (Cathode Ray Tubes). In the workplace, VDTs are mainly used as computer display terminals and may be found virtually everywhere, with greatest prevalence in the office setting.

Over the past decade or so, there has developed an ongoing controversy regarding the safety of VDTs from a number of standpoints, including their capacity to cause adverse pregnancy outcome, eye strain or damage, and even cancer. The debate and the research continue, but the current professional consensus tilts towards the opinion that properly designed and maintained VDTs do not pose a carcinogenic or teratogenic risk to workers. VDTs are, however, capable of causing significant eye strain, headaches, neckaches, and other somatic problems.

During the walk-through, the physician should observe whether VDTs are present and should inquire about their functionality and maintenance. Some resources recommend that, for continuous VDT activity, workers should be allowed rest periods of at least 10 to 15 minutes/hour, during which they can perform other activities.[78] Other resources offer no specific rest break recommendations or generalize that breaks should be taken informally, in an unscheduled manner, depending upon the intensity of work.[79,80] Problems with VDT use relate to many factors including:

- workstation design

- screen glare from ambient sources or outdoor light

- improper character/screen color combinations

- inadequate screen luminance

- lack of screen image stability and sharpness

- presence of display flicker

- improper positioning of written or typed text which the operator must read during transcription into the computer

- improper operator posture

As a corollary to the issue of the potential adverse health effects of VDTs, the physician should be aware that, over the past several years, significant problems have emerged with keyboard design and the subsequent claimed development of upper extremity problems in VDT operators. The main issue of keyboard design has related to keyboards with tall and abruptly angulated or sharp leading edges, which may force the operator to type with hands in extension and ulnar deviation, sometimes with the anterior surface of the wrist compressed against the keyboard itself. Medical problems claimed by keyboard operators have included carpal tunnel syndrome, thoracic outlet syndrome, and even reflex sympathetic dystrophy. Attempts to mitigate the ergonomic aspects of these problems have involved placing wrist rests in front of the keyboard leading edge or utilizing keyboards with slimmer profiles.

*Harmful Electromagnetic Frequencies*

*Ultraviolet (U-V) Radiation:* This form of electromagnetic radiation has wavelengths longer than x-rays but shorter than visible light. There are several schemes in physics by which the ultra-violet region is divided by wavelength.[81,82] The shorter wavelengths are the most energetic and have the greatest potential for adverse biological effect. The following definitions incorporate practical aspects of more than one scheme.

*Near U-V:* 3000-4000 Angstroms (1 Angstrom = $10^{-8}$ cm) or $3 \times 10^{-5}$ to $4 \times 10^{-5}$ cm. This is the so-called *black light* region.

*Far U-V:* 2000-3000 Angstroms or $2 \times 10^{-5}$ to $3 \times 10^{-5}$ cm.

*Extreme or Vacuum U-V:* 10-2000 Angstroms or $10^{-7}$ to $2 \times 10^{-5}$ cm.

From a photobiological standpoint, the U-V spectrum is also broken down into three regions.[82] The U-V region between 3150 and about 4000 Angstroms is called UV-A. The U-V region between 2800 and about 3150 Angstroms is called UV-B, the *sunburn* or *erythemal region.* UV-B radiation is also well absorbed by the cornea and represents the wavelengths that cause *welder's flash* or corneal keratitis.[83] The U-V region between 1000 and 2800 Angstroms is called UV-C. Wavelengths in this region have germicidal properties and are used in germicidal lamps for air and water purification. Some welding arcs also produce UV-C radiation. The U-V region between

1700 and 2200 Angstroms is the most efficient for production of ozone.[81]

*Ultraviolet TLVs*:   TLVs for U-V radiation are dependent upon wavelength, duration of exposure, and irradiance which is measured in units of milliwatts per square centimeter of biological surface.

Recall from note (VI-2) that dose is a function of concentration times duration of exposure.   The units of U-V dose, therefore, are (milliwatts x seconds)/cm² or millijoules/cm² where a joule is the metric unit of work in the MKS (Meter-Kilogram-Second) system. (The watt is the unit of power equal to one joule/s.)

In those cases where all factors are known, the TLV is expressed as a maximum dose (millijoules/cm²) to skin or eyes which should not be exceeded at any time during an 8-hour day for a specific U-V wavelength.   In principle, therefore, the TLV for U-V of a specific wavelength is a ceiling limit.

If the U-V source is *broadband* (white light or open arcs), the effective irradiance (EI) relative to a monochromatic source at 2700 Angstroms must be determined.   The units of EI can be expressed in milliwatts/cm². The TLV for *broadband* U-V exposure is expressed as the permissible exposure *time* allowed for a broadband U-V source of a specified effective irradiance.[84]

Note that the TLV for both types of U-V sources can be easily reconciled because when one multiplies the broadband EI level by permissible exposure time, the product can be shown to have units of millijoules/cm² which is  the same unit of dose as that for the maximum exposure level (ceiling) permitted for monochromatic U-V sources.

*The U-V Hazard*:  U-V radiation in the workplace (especially from arc-welding) can represent a significant hazard to the eyes.   In addition, because of its ability to initiate photochemical reactions, U-V radiation from arc welding, when present near chlorinated solvent tanks, can cause the generation of carbonyl chloride ($COCl_2$) or phosgene, an insidious gas (a chemical warfare favorite) which can cause delayed onset pulmonary edema (relatively mild warning properties; eventual hydrolysis to hydrochloric acid (HCl) in the lungs).

The breakdown of solvent into hazardous materials occurs in the U-V field around the arc, not in the arc itself. Any fugitive solvent vapors which enter the U-V field are subject to decomposition.[85] If you see unshielded arc welding taking place near uncovered chlorinated solvent tanks, therefore, consider the situation to be potentially dangerous.

*Microwaves (MW) and Radiofrequency (RF) Radiation: Definitions and TLVs:* MW and RF radiations are electromagnetic radiations which have many important uses in our modern society, mainly within the realms of communications and industrial production technologies.[86] MW radiation (technically, a subset or type of RF radiation), of course, is also commonly used in many homes and restaurants to prepare food. The primary, currently understood mechanism by which these types of radiation can produce harm to biological systems is secondary to their ability to heat tissue.

The OSHA standard (29 CFR 1910.97) for worker protection against MW and RF radiation applies to frequencies between 10 and 100,000 MHz (megaHertz or million cycles/s). This law has established a limit for occupational exposures not to exceed a power density of 10 milliwatts/$cm^2$ as averaged over a 0.1 hour (6 minute) period.[87]

The World Health Organization (WHO) considers its technical interest in MW and RF radiation to extend between frequencies of 0.1 and 300,000 MHz of which the RF range lies between 0.1 and 300 MHz (3 km to 1 m wavelength in air) and the MW range lies between 300 MHz and 300,000 MHz (1m to 1mm wavelength in air).[88] WHO indicates that permissible power densities over the entire MW/RF frequency range should be between 0.1 to 1 milliwatts/$cm^2$ and should offer a high enough safety factor to protect all workers continuously exposed throughout the workday (except, perhaps, pregnant women.)[89]   No specific standards for pregnant women are recommended except that WHO has noted that suspected adverse effects have been reported to occur at power densities exceeding 10 milliwatts/cm$^2$. [90]

The ACGIH has identified its domain of interest for MW and RF radiation and, therefore, its range of TLV applicability as extending between 0.03 MHz and 300,000 MHz.  Within this frequency range, the maximum power density permitted (TLV) varies according to frequency and is maximum at 100 milliwatts/cm$^2$ for the frequencies between 0.03 and 3 MHz.  Its minimum value of 1 milliwatt/cm$^2$ applies within several frequency ranges.  All TLVs are TWA for exposure periods of 0.1 hours or 6 minutes.[91]

The preceding, somewhat confusing and fairly frustrating paragraphs have been purposely included to alert the physician to the complexity of occupational standards, the lack of unanimity among equally respectable agencies, and the caution which must be exercised when discussing exposure data with workers, industrial hygienists, and management.  Remember, unless otherwise instructed, the OSHA standard is the legal one.  In a medical-legal forum, however, lower, more conservative exposure standards from other agencies may be invoked by one party or another (usually the plaintiff) as having legal importance in causation, based upon the standard's usefulness in establishing *state-of-the-art industrial health and safety practice* within a given industry or in a specific situation.

*Microwaves*:  Like U-V, permissible exposure limits for microwaves are dependent upon wavelength (or frequency), duration of exposure and power intensity which, as mentioned, is measured in units milliwatts per square centimeter of biological surface area.  Unlike U-V, shorter wavelength microwaves (higher frequency) are, generally, *less* dangerous to biological systems.  Microwave radiation greater than 3000 MHz in frequency, for example, is usually absorbed in the skin; but microwaves with frequencies less than 3000 MHz can penetrate and heat up underlying tissues.[86]

*RF or Radiofrequency*:  RF or radiofrequency radiation is utilized in a wide variety of industrial heating and sterilizing operations.  RF heaters have power outputs in the 200 kiloHertz (200,000 cycles/second) to the megaHertz range depending upon the particular type of industrial heating application.[92]   Whole body exposures to workers exposed to RF radiation are usually small, but there can be high localized magnetic flux densities and adverse temperature effects to the hands, extremities, or even the head of an operator, who is using inadequately shielded equipment.

*Lasers*:  *LASER* is an acronym for *Light Amplification by Stimulated Emission of Radiation.*  Lasers are used for a bewildering variety of industrial and medical processes and procedures.  The main target organ of the laser is the retina of the eye.  Even small lasers can be very dangerous to the retina if their power is focused to a very small area, thus increasing their power intensity.  The PEL or TLV for lasers is expressed in either millijoules/cm$^2$ (one joule is the basic unit of work in the MKS metric system) or milliwatts/cm$^2$ for a given laser wavelength and a specified exposure time.[93,94]

During the walk-through, the physician's main concerns should be that lasers are appropriately shielded and that fail-safe procedures and devices are in place to prevent the operator from looking directly into the beam or sustaining direct skin exposure.

Finally, it is important for the physician conducting the walk-through to note whether all potential, hazardous frequency exposure areas are properly designated with warning signs.

## (VI-10) - Ventilation

Ventilation is an extremely complex subject, and the physician must rely extensively upon the industrial hygienist for technical assistance in this regard.  A few principles which can assist the physician, however, are:

■ Observe the worker's activities and the movement of his or her breathing zone in space and time.

■ Locate the anticipated pathway of hazard movement and determine whether the hazard flows into or past the worker's breathing zone.  For example, if someone is performing bench soldering and a suction fan is located behind the worker's head, the fume may pass the worker's mouth and nose in greater amounts than it would if no ventilation were present.  Another example would be a situation in which a worker leans into a hood in which there is a suction fan overhead.  In this case, hazardous vapors will pass his breathing zone as they are drawn upward.  If, on the other hand, the hood is a *laminar flow* type in which highly filtered (HEPA) air flows *from* the top to the bottom of the enclosure and is evacuated through a slot in the table, then, the worker may not have a problem, because fresh air is passing the breathing zone.

■ Look for areas of ventilation stagnation, where a build-up of hazardous emissions could occur.

■ Inquire whether the ventilation present during the walk-through is capable of handling worst case conditions.  Also, remember that some employers cut down on outside make-up air during colder months because of the cost of heating it.

■ Remember to inquire about and to inspect the intake areas of general ventilation systems.  A possible source of contaminated air inside the plant can come from *outside* air, if make-up air intakes are located, for example, near a street level loading dock, where delivery trucks may idle while unloading. Ventilation system air intake ducts may also be located on the roof, downwind from a chimney on a neighboring building or even downwind from a waste emissions stack emanating from the plant itself.  In all three of the examples mentioned above, the contaminated air which enters the fresh air intake is called a *fugitive emission*, which may be overlooked as an important cause of workplace exposure problems.

- Be cognizant of the potential for *tight-building* problems and refer to reliable resources on this subject.[95-99]

- Remember that provision of adequate ventilation should be a much higher priority to the employer than issuing a respirator to the worker.

- Leave most of the mathematical formulae to the industrial hygienist, but one equation to consider tackling with a little technical back-up is provided by the ACGIH[100] and is nicely illustrated by McDermott.[101] This equation can be utilized when one wishes to calculate the amount of dilutional airflow ($Q_D$), which is needed to maintain the airborne contaminant level of a liquid (e.g., a solvent) at the TLV or PEL, given information on the amount of solvent *used up* per unit time and given a number of other factors are specified, as below:

$$Q_D = [L_R \times CF \times (\text{sp gr}) \times SF \times 10^6]/(M_r \times A_L) \quad \text{EQ[VI-13]}$$

where:

$Q_D$ = amount of dilutional airflow needed to achieve $A_L(\text{ft}^3/\text{min.})$

$L_R$ = amount of solvent or liquid used per unit time (e.g., pints/hr., gallons/hr., etc.)

CF = a conversion factor based upon the units of liquid consumption rate, $L_R$ (see Table VI-1)

sp gr = specific gravity[102] of the liquid or solvent relative to water = 1.0

SF = a dimensionless *safety factor* which, since greater than 1, will increase $Q_D$ to a larger value and will help take into account non-ideal ventilation situations. SF normally ranges from about 3 to 10 and is increased to the higher limits of this range when facility design does not permit good airflow and when larger numbers of workers are exposed

$M_r$ = the molecular weight of the solvent or liquid contaminant (dimensionless)

$A_L$ = the TWA acceptable airborne exposure limit, for example, PEL or TLV of the contaminant (ppm)

## (TABLE VI-1)

CONVERSION FACTORS (CF) FOR VARIOUS LIQUID CONSUMPTION RATES USED FOR CALCULATING DILUTIONAL AIRFLOW ($Q_D$) [101]

| Amount of Liquid or Solvent Used Per Unit of Time - $L_R$ | Conversion Factor CF |
|---|---|
| Pints/hr. | 6.7 |
| Pints/min. | 403.0 |
| Gallons/hr. | 53.7 |
| Gallons/min. | 3222.0 |
| Liters/hr. | 14.1 |
| Liters/min. | 846.0 |

*EXAMPLE*:

Assume that you wish to calculate the value of $Q_D$ in a workplace in which workers are *using up* a solution of isopropyl alcohol (IPA) (80% alcohol; 20% water) to clean table tops in an assembly area at the rate of 6 pints/hour.  The assembly area is not open enough to allow general dilutional ventilation to pass over all table tops with assurance, and about 10 workers are involved in the cleaning process.

Question: *What dilutional airflow is needed to keep the airborne levels of isopropyl alcohol (IPA) at levels acceptable to OSHA?*

The dilution ventilation equation can be utilized as follows:

- Since the IPA is 80% pure and 6 pints/hr. of solution are consumed (and presumably evaporated into the air), (0.8)(6 pints/hr.) or 4.8 pints/hr. of pure IPA are consumed. Thus: $L_R$ = 4.8 pints/hr.

- According to Table VI-10/1, for the units of pints/hr., CF = 6.7.

- The specific gravity of IPA[103] is 0.785 at 20°C. Thus, sp gr = 0.785.

- Given the layout of the assembly area and the number of workers involved, a safety factor of about 6 is estimated to be necessary. Thus, SF = 6.

- The molecular weight of IPA[103] is 60.10. Thus, $M_r$ = 60.10.

- The OSHA PEL for IPA (1990) is 400 ppm. Thus, $A_L$ = 400.

Substituting into EQ[VI-13]:

$$Q_D = (4.8 \times 6.7 \times 0.785 \times 6 \times 10^6)/[(60.10) \times (400)]$$
$$Q_D = (1.515 \times 10^8)/(2.404 \times 10^4)$$
$$Q_D = 6.3 \times 10^3 \text{ or } 6300 \text{ ft}^3/\text{min.}$$

The equation for $Q_D$ can also be used to solve for an estimated airborne concentration, if $Q_D$ is known, if the safety factor SF is set equal to 1.0, and if all other parameters are specified. Finally, one should be cautioned that calculated values of $Q_D$, even assuming relatively high safety factors, may not prevent local high concentration exposure to the worker, if the worker's breathing zone is in the path of the contaminant flow.

### (VI-11) - Housekeeping

The housekeeping in a workplace relates to how neat, uncluttered and, generally, safe the workplace appears. Good housekeeping reflects the health and safety attitudes of both management and employees. In a well kept workplace, aisles are often marked with tape or paint, are unobstructed, and are not littered with objects which can cause *slips, trips, or falls*. Raw materials, products, and inventory are stored neatly and safely. Signs are clearly visible and unobstructed and, more often than not, even in production areas the workplace is clean. The brilliance of good housekeeping can sometimes hide underlying health and safety problems; but, as a general rule, good housekeeping practices reflect a strong health and safety program.

### (VI-12) - Sanitation

Observe whether employees are permitted to eat, drink, or smoke at their workstations and whether such activities may result in the ingestion or inhalation of hazardous materials. In almost all cases, these activities should be banned* from contaminated areas and confined to areas entirely separate from workstations.

Lunchrooms and eating facilities should be located near washrooms. Ideally, the worker should be as clean as possible before eating. To attain this goal, it is certainly necessary to wash up, and it may be necessary to change clothes before eating. Under circumstances in which the potential for contamination of personal clothing is high, there should be lockers and showers available which are well separated from contaminated areas. The facilities should permit the worker, upon arrival, to change from street clothes to work clothes and, upon departure for home, to shower before changing back into street clothes. Separate sanitary facilities and lockers should be available for both men and women in all cases.

---

* When this rule is disobeyed, the adverse consequences can be potentially serious. Consider the problem which may occur, for example, when an individual smokes a cigarette in the vicinity of a freon-containing vapor degreaser. The freon which passes through the lighted cigarette will partially combust, forming phosgene (see VI-9; U-V), among other harmful substances. Phosgene has the capacity to cause delayed onset pulmonary edema after being converted to hydrochloric acid in the lung.

**(VI-13) - Noise and Vibration**

*Sound Pressure, Sound Pressure Level and the dB*

Noise levels in the workplace are measured in units called *decibels*, or *dB*. The decibel is the unit of sound pressure level and is expressed by the formula:[104,105]

$$dB = 20 \log (p_1/p_2) \qquad\qquad EQ[VI\text{-}14]$$

where

$p_1$ = is the pressure of the sound being measured.

$p_2$ = a reference pressure equal to 0.0002 microbars or millionths of a bar (one bar is equal to approximately 760 mmHg or one atmosphere and is defined as $10^6$ dynes/cm$^2$ where the dyne is the metric unit of force in the CGS system).

log = common logarithm (base 10)

Assume that the ratio $(p_1/p_2)$ of measured sound pressure to the reference pressure is 1000, then sound pressure level in dB is (from EQ[VI-14]):

$dB = 20 \log (1000)$

$dB = 20 \times 3$

$dB = 60$

Next, assume that the ratio $(p_1/p_2)$ is doubled to 2000. Then

$dB = 20 \log (2000)$

$dB = 20 \times 3.3$

$dB = 66$

Note that a doubling of sound pressure resulted in an increase of only 6 dB in sound pressure level (this relationship will always be true).

Next, assume that the ratio $p_1/p_2$ = 1000 is multiplied by 10 to become 10,000. Then:

$$dB = 20 \log(10,000)$$
$$dB = 20 \times 4$$
$$dB = 80$$

Note that increasing sound pressure by an order of magnitude or by a factor of 10 results in an increase of 20 dB in sound pressure level (this relationship also will always be true).*

The important point to gather from these examples is that the decibel is derived from the *logarithm* of a ratio of sound pressures and *not* from the ratio of sound pressures alone.  Thus, relatively modest changes in dB level can only be achieved by very large changes in sound pressure ratio.  This mathematical phenomenon has some other fascinating ramifications.  At both high and low decibel levels, the doubling of sound pressure ratios increases the decibel level by only 6 dB.  When this doubling occurs at high dB levels, however, the change in actual sound pressure needed to create a 6 dB increase may be enormously greater than at lower dB levels.

*EXAMPLE*:

Assuming a sound pressure reference level ($p_2$) of 0.0002 microbar, a 20 dB sound level would correspond to a sound pressure ($p_1$) of 0.0020 microbars [i.e., 20 dB = 20 log (.0020/.0002) or 20 dB = 20 log(10) or 20 dB = 20(1)].

---

* It should be emphasized that the decibel can also be expressed in terms of the ratio of two *sound intensities* $S_1$ and $S_2$, each measured in watts/m$^2$, as dB = 10 log ($S_1/S_2$). From this equation one can easily show that a doubling of *sound intensity* will result in an increase of 3 dB in *sound pressure level*. Any 3 dB increase in sound pressure level, therefore, represents a doubling of sound intensity. By the same token, an increase in sound intensity by a factor of 10 will result in a 10 dB increase in sound pressure level, and an increase in sound intensity by a factor of 100 will result in a 20 dB increase in sound pressure level. The notion of changes in sound intensity must be differentiated from changes in sound pressure as described above. In the latter case, as already shown by equation (VI-14) and the text example, a doubling of sound pressure will result in a 6 dB increase in sound pressure level.

A 26 dB sound level would correspond to a sound pressure level ($p_1$) of 0.0040 microbars [i.e. 26 dB = 20 log (.0040/.0002) or 26 dB = 20 log 20 or 26 dB = 20 (1.3)].

The actual increase in sound pressure needed to increase sound pressure level from 20 to 26 dB is, therefore, (0.004 - 0.002) microbars or 0.002 microbars.

Now consider:

A 100 dB sound pressure level would correspond to a sound pressure ($p_1$) of 20 microbars [i.e. 100 dB = 20 log (20/.0002) or 100 dB = 20 log (100,000) or 100 dB = 20(5)].

A 106 dB sound pressure level would correspond to a sound pressure ($p_1$) of 40 microbars [i.e., 106 dB = 20 log(40/.0002) or 106 dB = 20 log (200,000) or 106 dB = 20 (5.3)].

The actual increase in sound pressure needed to boost sound pressure level from 100 to 106 dB is, therefore, (40-20) microbars or 20 microbars.

Now, if we compare the increase in sound pressure needed to boost sound pressure level from 20 to 26 dB to the increase needed to boost sound pressure level from 100 to 106 dB, we find that it took only 0.002 microbars in the former case and 20 microbars in the latter case! Stated in other terms, it took (20)/(0.002) or 10,000 times more of a sound pressure increase to raise the sound pressure level from 100 to 106 dB than it did to raise the sound pressure level from 20 to 26 dB (not withstanding the fact that both changes represented a doubling of sound pressure).

The above example should clearly illustrate, for the physician (from the point of view of the eardrum), how much more serious a jump of 6 dB in sound pressure level is at, say, 80 dB than at 90 or 100 dB (notwithstanding the eardrum's ability to accommodate rapid changes in noise levels via the tensor tympani muscle reflex)!   Given, moreover, that 1 dB approaches the smallest change in sound pressure level which can be appreciated by the human ear,[106] even such a small change at high dB levels must be accompanied by much larger increases in sound pressure than at lower dB levels.

*Sound Pressure Level Versus Sound Level*

The human ear senses *loudness* based not only upon the *sound pressure* but also upon the frequency or pitch of the sound.   In addition, the manner in which loudness varies with frequency also depends upon sound pressure.   If a sound measuring instrument is designed to take these factors into account when measuring broad-frequency sound or noise, it will automatically weight or sense certain frequencies more than others.   In other words, it will *listen* to the noise in a manner more like the human ear does and less like an ordinary microphone which will *sense* the actual frequency dependent sound pressure levels.   A weighted sound measuring instrument is called a *sound level meter* and the sound levels it measures are called *A, B, or C-weighted levels* (n.b. A weighting most closely resembles human hearing).   A microphone system which is completely *unweighted* and shows no recording bias at any frequency is called a *sound pressure level meter*.

For industrial hygiene work, the A weighted instruments are used and the frequency weighted sound level readout from the instrument is measured in dB(A).   The physician should realize that A weighted instruments become relatively *deaf* as frequencies drop below 200 Hz, *hear better* than sound pressure level microphones between 1000 and 5000 Hz, and then begin to *lose their hearing* again between 5000 and 10,000 Hz.   C and B weighted instruments *hear* a little better at lower frequencies than A weighted instruments (C better than B) and are equally more *hard of hearing* than A-weighted instruments at frequencies greater than about 1000 Hz.[107]

*Combined Noise Level from Two Sources*

Given the logarithmic nature of the dB or dB(A) scale, the physician should realize that one cannot simply add together the noise levels from two different machines. References are available to assist you in combining noise levels from two sources when the sources are relatively close together and when you can measure the noise level of each source separately.[108]   Table (VI-2) is derived from a noise level combination graph.

*EXAMPLE*:

Source #1:  dB(A) = 67

Source #2:  dB(A) = 72

[dB(A) of source #2] - [dB(A) of source #2] = 5 dB(A)

From Table (VI-2), the amount to be added in dB (A) to the higher level source when the difference between the sources is 5 dB(A) is 1.2 dB.    Thus, the combined noise level of both sources is 72 + 1.2 = 73.2 dB(A).

The physician should realize that workers are permitted exposure to specific industrial noise levels (TLVs) depending upon the duration of exposure.   For example, according to the ACGIH,[109] the worker can be exposed to 85 dB(A) for 8 hours; 90 dB(A) for 4 hours; 95 dB(A) for 2 hours; 100 dB(A) for 1 hour; 105 dB(A) for 1/2 hour; 110 dB(A) for 1/4 hour and 115 dB(A) for 1/8 hour per day. No exposure to continuous or intermittent noise is permitted in excess of 115 dB(A).

*Impulse or Impact Noise*

There are also ACGIH guidelines provided for *impulse* or *impact noise* (e.g., hammer drill, jackhammer, stamping machine, etc.) which specify the permitted number of impulses or impacts per day as a function of peak dB sound pressure level [not dB(A)!].[109]

## TABLE (VI-2)

## TABLE FOR ESTIMATING THE COMBINED NOISE LEVEL OF TWO SOURCES

(From[108] - originally appeared in the form of a noise level combination graph in Sataloff, J. *Industrial Deafness*, Blakiston Division, McGraw-Hill, New York, 1957)

| Difference in dB(A) between the two sources whose noise levels are being combined | Increment in dB(A) to be added to the higher level source to derive the combined noise level of both sources |
|---|---|
| 0 | 3.0 |
| 0.5 | 2.8 |
| 1.0 | 2.6 |
| 1.5 | 2.4 |
| 2.0 | 2.1 |
| 2.5 | 1.9 |
| 3.0 | 1.8 |
| 3.5 | 1.6 |
| 4.0 | 1.5 |
| 4.5 | 1.3 |
| 5.0 | 1.2 |
| 5.5 | 1.1 |
| 6.0 | 1.0 |
| 6.5 | 0.9 |
| 7.0 | 0.8 |
| 7.5 | 0.7 |
| 8.0 - 8.5 | 0.6 |
| 9.0 - 9.5 | 0.5 |
| 10.0 - 11.0 | 0.4 |
| 11.5 - 12.5 | 0.3 |
| 13.0 - 14.5 | 0.2 |
| 15.0 | 0.1 |

*Vibration*

Vibration in the workplace can occur from many sources and can adversely affect the fingers, upper and lower extremities, or even the whole body. There are several excellent references entirely dedicated to this subject.[110-113] During the walk-through, look for sources of vibration which may include electrical hand tools like sanders, drills or rivet guns, or large pneumatic devices like jackhammers or gasoline-powered devices such as chainsaws.

Vibration measurements must be made by a qualified industrial hygienist. Measurements are made using accelerometers in three mutually orthogonal directions at a point nearest to that in which the vibration enters, for example, the hand. ACGIH recommended limits are measured in $g_e$ forces where $g_e$ is the acceleration of gravity equal to 9.81 meters/s$^2$. The physician should realize that excessive vibration exposure to the hands and fingers, in particular, can lead to the *Vibration White Finger Syndrome*, a disorder similar to Raynaud's Phenomenon. Excessive vibration has also been implicated in the development of other disorders including carpal tunnel syndrome.

The physician should be observant for steps taken by workers and management to mitigate vibration exposure, with such protective measures as vibration attenuating gloves, the use of antivibration foam materials on the handles of vibrating tools, and the avoidance of using vibrating tools under cold stress conditions.[114]

## (VI-14) - Signs and Labels

Adequate and clear signage in the workplace is absolutely essential for maintaining a safe and healthy working environment. The task of performing this aspect of the walk-through can be made easier, if the physician conceptualizes the various types of signs and the purpose of each.

Remember, the purpose of signs is to communicate information and/or to communicate imperatives, namely directions or warnings to do something or not to do something. In some cases, signs must be multilingual in this regard.  Signs must be large enough and clear enough to read, must be placed conspicuously, and must be unobstructed.  Consider how important these factors are not only for workers and visitors, but also for firefighters or ambulance personnel who may have to respond to a plant disaster.

There are several OSHA regulations which deal directly with sign design, color scheme, location and size (e.g., 29 CFR parts 1910.36, 1910.144, 1910.145, 1926.200 and 1917.128) and with which the plant safety manager and/or industrial hygienist should be familiar.  Your role is to inquire about sign policy and to judge for yourself whether the policy seems to be enforced and whether signage, in your opinion, appears adequate.  As you walk through the workplace, look for signs related to:

- the *location of critical areas* like exits, walkways, locker rooms and bathrooms, eating areas, the first aid room, confined spaces, chemical storage rooms, etc.

- the *location of critical items* such as first aid kits, fire extinguishers, telephones, self-contained breathing apparatus (SCBA), emergency eye wash and shower stands, etc.

- the *provision of instructions and warnings* such as:  the specification of personal protection requirements (e.g., hard hats, safety glasses, ear protection, etc.), *Do Not Enter* signs, specific hazard warning signs (e.g., biohazards, chemical hazards, noise, high voltage ionizing radiation, lasers, ultraviolet, microwaves, etc.), weight limit signs, no smoking signs, out-of-order signs, traffic signs, fork-lift direction signs in warehouses, etc.

Remember, as stated in other sections of this appendix, your role is not to perform an OSHA compliance audit but, instead, is to provide occupational medical feedback to the employer and to ask questions which will inspire the employer to think about his or her formal compliance obligations.

## (VI-15) - Chemical Containers

Precariously positioned, improperly sealed, and physically compromised chemical containers in the workplace can present serious health, safety, and fire hazards. In addition, unsecured compressed gas cylinders not only can fall and seriously injure a worker but also can sustain valve damage and release their contents in an uncontrolled manner.

It is not uncommon to walk through some workplaces and see open pools of solvent within unventilated degreasing tanks or open pans of similar volatile liquids on work benches. Often, even the commercial container in which a chemical is sold will be left with its lid off. Clearly, these situations enhance evaporation and convective losses, causing airborne concentrations of chemical hazards to increase. Recommend to the employer that workers *KEEP THE LIDS ON* all such containers and cover up sources of potential contact between room air and liquid chemical hazards.

Some workplaces are seemingly littered with numerous, partially used chemical containers of indeterminate age. Virtually all containers will eventually lose their structural integrity and begin to leak over time. Such leaks may be dangerous or even explosive. If you observe containers in this condition, inform your escort immediately.

Finally, observe whether chemical containers are located in areas where they may be in physical jeopardy. It is not unusual, for example, to see containers, balanced precariously on a shelf or left in an aisleway.

## (VI-16) - Chemical Storage

Chemicals should be stored in approved chemical storage cabinets or rooms which are *adequately ventilated*. So often, the physician may see chemicals stored in office supply or filing cabinets, refrigerators, or other unsuitable places. Employers may utilize these substandard options because approved chemical storage cabinets are very expensive.

Inquire whether incompatible chemicals are stored together (e.g., acids and bases; calcium hypochlorite and hydrochloric acid or acetone and strong oxidizers such as permanganate salts, etc.) and ask whether Material Safety Data Sheets were consulted in conjunction with selecting storage locations.

In some cases, chemicals are moved in small containers or are piped from large bulk storage facilities. Determine whether these containers are adequately labeled and are appropriate for storing the material being conveyed. If liquids are piped to a remote location, ask whether the plumbing and the distribution spigots are properly labeled. Don't be surprised to see temporary, makeshift signs which identify critically important chemical distribution spigots. There are usually no problems until a new employee or a worker from another section of the plant mistakes such a spigot for a water faucet!

## (VI-17) - Confined Spaces

Confined spaces are one of the greatest *potential killers* in the workplace. The federal government is well aware of the dangers of confined spaces and has published several important documents, outlining the seriousness of the hazard and guidelines to mitigate harm.[115-117] The confined space is defined by NIOSH as follows:

> ...a space which by design has limited openings for entry and exit; unfavorable natural ventilation which could contain or produce dangerous air contaminants, and which is not intended for continuous employee occupancy. Confined spaces include but are not limited to storage tanks, compartments of ships, process vessels, pits, silos, vats, degreasers, reaction vessels, boilers, ventilation and exhaust ducts, sewers, tunnels, underground utility vaults, and pipelines.[118]

NIOSH also defines subclasses of confined spaces as follows:

| | |
|---|---|
| Confined Space, Class *A* | A confined space that presents a situation that is immediately dangerous to life or health (IDLH). These include but are not limited to oxygen deficiency, explosive or flammable atmospheres, and/or concentrations of toxic substances. |
| Confined Space, Class *B* | A confined space that has the potential for causing injury and illness, if preventive measures are not used, but is not immediately dangerous to life and health. |
| Confined Space, Class *C* | A confined space in which the potential hazard would not require any special modification of the work procedure.[119] |

Inquire whether the employer has identified confined spaces and has developed standard operating procedures for entry. Entry into a confined space is usually allowed only by a written permit system in which the space is tagged and entry is not permitted until appropriate testing of the environment has been completed for at least oxygen concentration and the presence of a combustible atmosphere. Workers who enter confined spaces should use life lines (to assist in rescue) and air-line (external air supplied) positive pressure respirators (See VI-8).

## (VI-18) - Emergency Equipment

As discussed in Section (VI-14), all emergency equipment must be well marked and easy to locate. The condition of this equipment should be audited on a regular basis and a written inspection record should be attached to all items. The emergency eyewash and shower stands should be tested regularly as well. In this regard, the eyewash faucet heads (which are positioned in an upward facing orientation whereby they can collect dust and debris on the aeration screen) should be covered at all times when not in use. First aid kits (see VI-19) must be properly maintained and workers should be discouraged from taking more items than they need (i.e., - for car or home use!). SCBA air tanks must be kept full at all times and broken down, cleaned and inspected carefully after each use. Safety harnesses, life lines and other rescue equipment must be stored neatly in an untangled manner, so that it is ready at a moment's notice.

## (VI-19) - First Aid Kits

Federal law requires that first aid kit contents must be approved by a consulting physician (29 CFR 1910.151). Most employers comply with this requirement by buying prepackaged or customized first aid kits from a safety equipment or product company which stocks kits approved by their own physician or by a consulting physician to the manufacturer of the kit. This process is fine as long as you agree with the contents of the kit. Some approved kits contain an unbelievable assortment of items which are available to the employee without any medical supervision. Here are a few recommendations about first aid kits:

- Kits which are not medically supervised by at least an LPN should contain *no oral medication of any kind*. Topical antibiotic ointment or typical first aid creams are fine. The provision of aspirin, acetaminophen, decongestants, antihistamines, non-steroidal antiinflammatory agents, antacids and combination preparations is a big business for the supplier. You may find, therefore, that the well intentioned employer has been sold a small pharmacy by his friendly sales representative. The unsupervised provision of any of these medicines to a worker for a self-diagnosed problem, however, can create serious worker compensation liability for the employer and can sometimes cause significant harm to the employee (e.g., aspirin or non-steroidal anti-inflammatory agents self-administered for treatment of chronic abdominal pain; antihistamines self-administered to treat cold symptoms in a forklift operator or steel worker who works at heights, etc.). If oral medications are available at work, they must be dispensed by an MD or a nurse who is working under a *written* medical protocol provided by an MD. Logs and inventory sheets must be maintained.

- Do not stock any chemical neutralizing agents in the first aid kit. For local, minor eye irrigation (not requiring an eyewash) use only sterile normal saline. Any agent which is capable of chemically neutralizing a foreign substance in the eye may generate heat in the process and cause further damage to lids, cornea, sclera, conjunctiva or lacrimal apparatus.

- Suggest that the employer routinely inventory the first aid kits to be certain that supplies are adequate for justifiable utilization and to avoid the dwindling of supplies at an unexplained pace.

- Be certain that kits are kept as clean as possible and contain a diverse and useful array of bandages, tape, scissors, irrigation fluids, topical antibacterial preparations and clean, absorbent materials to control bleeding (sanitary napkins are excellent additions to the kit for this purpose).

- Be certain that you are professionally satisfied with the quality and contents of the kit before voicing any approval. Chances are good that you will customize the kit in some manner to meet the specific needs of the work environment.

## (VI-20) - Materials Handling

Observe whether the workers are equipped with the proper tools which they need to move, pour, cut, bend, open, close, stack, store, or lift materials in the workplace. It is not uncommon, for example, to see an individual attempt to place a large, heavy box on an overhead shelf which is clearly out of reach, while teetering on a chair or stool just before falling. If the worker doesn't fall, he may sustain a cervical or lumbar strain.

Workers must have available and must be forced to use safe ladders, lifting assistance devices and other necessary tools for materials handling. Even when equipped with the proper tools, workers must not be asked to undertake tasks for which they are not physically capable or in which they may risk injury by virtue of the method or performance pace of job activities. Observe workers while they perform their normal duties whenever possible and make recommendations accordingly.

## (VI-21) - Employee Behavior

Your job during the walk-through is not to be a policeman, but it is useful for you to convey your perceptions to the employer about discrepancies between stated policies and worker or management compliance as seen during the walk-through. If you see a pop can or half eaten sandwich on a soldering bench, mention it to your escort. If your escort seems unconcerned, then relay the observation to your escort's supervisor. Your job on the walk-through is to be observant, objective, creative and helpful. Sometimes, however, the latter requires extreme candor and a little risk-taking.

## (VI-22) - Physician Self-Reporting

If you, personally, experience an adverse reaction, for example, to an odor or irritant during the walk-through or if you almost trip over a compressor hose and break your neck, tell your escort about the problem immediately. If you do not experience any problems in chemical hazard areas, remember that your non-reaction is no assurance that problems do not exist.   Only appropriate industrial hygiene monitoring can resolve concerns.

## (VI-23) - Immediate Dangers to Life or Health

If, during the walk-through, you observe conditions that are *very* serious and require immediate mitigation, inform your escort and insist upon swift action.   If your concerns and perceptions are confirmed, meet with the individual who can make the decision to correct the problem.  Document such concerns in writing and provide them to your client.  Expect that employers will respond appropriately in virtually all cases, but follow up with your client to make certain that corrective actions have been taken.

If you are ignored or have reason to believe that no corrective action is or will be taken for a problem that is immediately dangerous to life or health, it should not be difficult for you, as a physician, to determine your proper course of immediate action.

# GLOSSARY

**AAD**: Ambient Air Dependent (respirator) (see VI-8).

**ACGIH**: American Conference of Governmental Industrial Hygienists.

**ADA**: Americans with Disabilities Act.

**Additive effect**: An effect which may be added to other effects when determining overall effect.

$A_L$: A TWA expressed in ppm.

**AL**: Action level (see VI-2).

**Angstrom**: $10^{-10}$ meters.

**apostilb**: The luminance of a perfectly diffuse, perfectly reflecting surface, illuminated by an illuminance of one lux (see VI-9).

$A_r$: IUPAC symbol for atomic weight.

**ASA**: American Standard Abbreviation.[120]

**asb**: apostilb.

**atm**: ASA symbol for atmosphere.

**Atmosphere**: A unit of air pressure equal to approximately 760 mmHg or 1.01325 bars at sea level and 15°C.

**Atomic weight**: The relative weight of an atom compared to $^{12}C = 12$. For a pure isotope, the atomic weight rounded off to the nearest integer represents the total number of neutrons and protons, making up the atomic nucleus.[121]

**Avogadro's number**: The number of molecules or atoms in one mole of a compound or element, respectively; about $6.023 \times 10^{23}$.

**bar**: International unit of pressure equal to $10$ N/cm$^2$, to approximately $0.987$ atm ($750.12$ mmHg at sea level and $15°C$) and to $10^5$ pascals.

**BEI**: Biological Exposure Index (see VI-2).

**Biological availability**: Relates to whether a hazard is capable of entering the body of an exposed individual or animal, by whatever route, such as inhalation, ingestion, or skin absorption.

**Biological monitoring**: The practice of collecting expired air, blood, waste, saliva, hair, or other tissue or fluids in order to determine the presence of a contaminant or its metabolite and in order to correlate analysis results with any exposure level.

**Break-through**: The phenomenon in which a gaseous or particulate contaminant passes through a filtering or adsorbing medium because of reduced collection efficiency of the medium.

**Breathing zone**: That space near the body from which an individual acquires inspired air.

**BSD**: *Best's Safety Directory*.

**c**: IUPAC symbol for the speed of light ($3 \times 10^{10}$ cm/s).

**C**: Concentration (units vary; often in mg/m$^3$).

**candela:** SI unit of light intensity or luminous intensity emitted from approximately 0.01667 cm$^2$ of the surface of a black body radiator at the freezing or solidification point of platinum, (2040°K) under one atmosphere of pressure.

**candela/m$^2$:** SI derived unit of luminance.

**cd:** SI symbol for candela.

**CF:** Conversion factor based upon the units of liquid consumption rate.

**CGS:** A coherent system of metric measurement units based upon the centimeter, gram, and second.[122]

**Chemical mixtures:** See VI-2.

**cm:** CGS symbol for the centimeter.

**$C_m$:** The number of cd/m$^2$.

**CRT:** Cathode Ray Tube - (see VI-9).

**$C_t$:** Contrast, a unitless term indicating the fractional amount by which a brighter object is brighter than a darker one (EQ[VI-12]).

**D:** Absorbed dose (e.g., mg or microgram).

**dB:** Decibel or sound pressure level (see VI-8 and VI-13).

**dB(A):** *A* weighted sound pressure level, derived from mathematical or electronic modification which renders recorded sound pressure levels (as a function of noise frequency) more like human hearing than unweighted recordings.

**DBT:** Dry bulb temperature (see VI-6).

**Demand Airline Device:**  An EAS respirator which provides breathing air when negative pressure is created by the act of inhalation (see VI-8).

**Dose:**  The overall amount of a substance taken into the body by whatever means.

**dyn:**  CGS symbol for the dyne.

**dyne:**  CGS derived unit of force equal to one $(g)(cm)/s^2$.

**EAS:**  External Air Supplied (respirator) (see VI-8).

**EI:**  Effect irradiance - relates to broadband U-V sources measured in $mW/cm^2$.

**EPA:**  Environmental Protection Agency.

**erg:**  CGS derived unit of work equal to one (dyn)(cm).

**Ergonomics:**  The scientific discipline dedicated to the study of the interrelationships between human workers and their physical work environment (see VI-4).

**Excursion limit:**  (see VI-2).

**f:**  Frequency of electromagnetic radiation.

**foot-candle:**  A unit of illuminance defined as one $lm/ft^2$ (see VI-9).

**foot-Lambert:**  The luminance of a perfectly diffuse, perfectly reflecting surface, illuminated by an illuminance of one foot-candle (see VI-9).

**ft:**  ASA symbol for the foot.

**ft-c**:  ASA symbol for the foot-candle.

**ft-L**:  ASA symbol for the foot-Lambert.

**Fugitive emission**:  An unwanted quantity of contaminated air which originates from an often unsuspected source, usually in relatively close proximity or upwind from the fresh air intake of a ventilation system.

**Functional operation**:  A workplace activity which usually involves work within more than one unit operation, is dynamic in place (i.e., is not fixed), and is usually *not* repetitive.

**g**:  CGS symbol for the gram.

**$g_e$**:  The acceleration of gravity equal to 9.81 m/s$^2$.

**$GM_r$**:  Gram molecular weight.

**Gram atomic weight**:  The weight of one mole of an element in grams, numerically equivalent to its atomic weight; also equal to the weight in grams of Avogadro's number of atoms of the element.

**Gram molecular weight**:  The weight of one mole of a pure compound in grams; numerically equivalent to its molecular weight; also equal to the weight in grams of Avogadro's number of molecules.

**GT**:  Globe temperature (see VI-6).

**Hazard** - See (VI-1).

**HEPA filter**:  High Efficiency Particulate Air filter.

**Hertz**:  SI derived unit of frequency equal to one cycle per second.

**HHE:** NIOSH nomenclature for a Health Hazard Evaluation.

**Housekeeping:** See (VI-11).

**Humidity, relative:** The percentage of moisture in the air at a given temperature relative to the amount of moisture the air could hold if fully saturated at the same temperature (see VI-6).

**Hygrometer:** See psychrometer.

**Hz:** SI symbol for Hertz.

**I:** Illuminance.

**IDLH:** Immediately Dangerous to Life or Health (see VI-2).

**Illuminance:** The illumination of a surface measured in lux or foot-candles.

**Impulse or impact noise:** Short duration, repetitive noise.

**Independent effect:** An effect which is entirely unrelated to other effects and under a particular exposure scenario would not be additive to or synergistic with other effects.

**IPA:** Isopropyl alcohol.

**IUPAC:** International Union of Pure and Applied Chemistry.[123]

**J:** SI symbol for the joule.

**joule:** SI derived unit of work equal to one (N)(m).

**K:** Absorption percentage.

**kg:** SI symbol for the kilogram.

**km**:  ASA symbol for kilometer.

**l**:  ASA symbol for the liter.

**L**:  Luminance.

**LASER**:  The acronym for *Light Amplification by Stimulated Emission of Radiation*.

**L$_{brighter}$**:  Luminance of the brighter object.

**L$_{darker}$**:  Luminance of the darker object.

**LF**:  Luminance factor.

**lm**:  SI derived symbol for luminous flux or lumen.

**log**:  An exponent, relative to a given base (in this case 10), such that when the base of the log is raised to the power of the log of the number, the result will be the number (e.g., if the number is 5 and the base of the log is 10, then $10^{\log 5} = 5$).

**LR**:  Luminance ratio.

**L$_R$**:  Amount of solvent or liquid used per unit time.

**Lumen**:  Unit of luminous flux emitted from a point source of one candela intensity through a solid angle of one steradian (see VI-9).

**Luminance**:  The *brightness* of a surface, measured in $cd/m^2$ (see VI-9).

**Luminance factor**:  The ratio of the luminance of a reflecting surface, viewed in a given direction, to that of a perfectly white, perfectly diffusing surface identically illuminated.[75]  (see VI-9).

**Luminance flux:**  The *flow rate* of light measured in lumens (see VI-9).

**Luminance ratio:**  The ratio between the luminance of a task and the luminance of the background or the surface upon which the task is being performed.[77]

**lux:**  A unit of illuminance defined as one $lm/m^2$.

**lx:**  SI symbol for illuminance or lux.

**m:**  SI symbol for the meter.

**MCL:**  Maximum concentration level:  The maximum concentration of a contaminant permitted in drinking water as stipulated by the U.S. Environmental Protection Agency.  Levels are set under the authority of the Safe Drinking Water Act of 1974 (PL 93-253).

**mg:**  A milligram or $10^{-3}$ grams.

**microbar:**  $10^{-6}$ bar.

**micron:**  $10^{-6}$ meters.

**Microwave:**  Electromagnetic radiation within the radiofrequency range between wavelengths of about 1mm and 1m.

**MKS:**  A coherent system of metric measurement units based upon the meter, kilogram, and second.

**mm:**  IUPAC/SI symbol for the millimeter.

**mmHg:**  Millimeters of mercury (pressure).

**mol:**  SI symbol for the mole.

**mole**:  The amount of a compound or element equal to Avogadro's number of molecules or atoms, respectively.

**Molecular weight**:  The sum of the atomic weights of the total number of individual atoms in a molecule.

**$M_r$**:  IUPAC symbol for molecular weight.

**MSDS**:  Material Safety Data Sheet.

**MW**:  Microwave.

**mW**:  IUPAC/SI derived symbol for the milliwatt.

**N**:  SI symbol for the newton.

**$N_A$**:  IUPAC symbol for Avogadro's number.

**newton**:  SI derived unit of force equal to one $(kg)(m)/s^2$.

**NIOSH**:  National Institute of Occupational Safety and Health.

**nit**:  A unit of luminance equal to one $cd/m^2$.

**Non-respirable particulate**:  A particulate unable (usually too large) to enter any part of the lung.

**OSHA**:  Occupational Safety and Health Administration.

**P**:  IUPAC symbol for pressure.

**Pa**:  SI symbol for the pascal.

**Particulate**:  A small particle.

**Particulate deposition**:  The deposition of particulates, sometimes in the lung or nasopharynx or on the skin.

**pascal**:  The SI derived unit for pressure equal to one $N/m^2$ or $10^{-5}$ bar.

**PATHMAX$^{TM}$**:  Parametric Approach to Health Maximization (see Appendix I).

**PEL**:  Permissible Exposure Limit (see VI-2).

**Potentiated effect**:  An effect which when added to other effects will create an overall effect which is greater than the sum of individual effects.

**ppm**:  Parts per million (usually by volume).

**Psychrometer**:  An instrument, also known as an hygrometer, used to measure relative humidity (see VI-6).

**$Q_D$**:  Dilutional airflow.

**r**:  IUPAC symbol for radius.

**rad**:  SI symbol for the radian.

**Radian**:  The central angle of a circle which subtends a length of arc equal to the radius of the circle (about 57.3°).

**Reflectance**:  A dimensionless quantity which represents the fraction of illuminance which is reflected as luminance from a surface.

**REL**:  Recommended Exposure Limit (see VI-2).

**Respirable particulate**:  A particulate capable of entering the lung.

**RF**:  Radio Frequency radiation (see VI-9).

**RH**:  Relative Humidity.

**s**:  SI symbol for the second.

**S**:  Sound intensity measured in $W/m^2$.

**sb**:  CGS symbol for the stilb.

**SCBA**:  Self-Contained Breathing Apparatus (see VI-8).

**SF**:  A dimensionless safety factor (see VI-10).

**SI**:  International System of Units. [124,125]

**Skin** (*skin*):  A TLV notation relating to skin exposure (see VI-2).

**Sound level**:  Noise level measured in dB(A).

**Sound level meter**:  An instrument that has the capability of measuring sound in dB(A).

**Sound pressure level**:  Noise level measured in dB.

**Sound pressure level meter**:  A microphone, used in conjunction with other equipment, that can measure sound pressure levels (dB).

**sp gr**:  ASA symbol for specific gravity or the ratio of the density of a substance to the density of a reference substance (dimensionless)[,102] usually water.

**sr**:  SI symbol for the steradian.[123,124]

**STEL**:  Short-Term Exposure Limit (see VI-2).

**steradian**:  The solid angle of a sphere with radius, r, which subtends a surface area on the sphere equal to $r^2$.

**stilb**:  CGS derived unit for luminance equal to one $cd/cm^2$.

**Stress**:  (See VI-5).

**Synergistic effect**:  See:  Potentiated effect.

**t**:  IUPAC symbol for time.

**TLV**:  ACGIH abbreviation for Threshold Limit Value (see VI-2).

**TLV-C**:  ACGIH abbreviation for Threshold Limit Value - Ceiling (see VI-2).

**TLV-TWA**:  ACGIH abbreviation for Threshold Limit Value - Time-Weighted Average (see VI-2).

**TWA**:  ACGIH abbreviation for Time-Weighted Average (see VI-2).

**Unit operation**:  A workplace activity which is usually static or fixed in place and repetitive.

**U-V**:  Ultra-violet (see VI-9).

**UV-A**:  U-V region between about 3150 and 4000 Angstroms.

**UV-B**:  U-V region between about 2800 and 3150 Angstroms - the sunburn, erythemal, or welder's flash region.

**UV-C**:  U-V region between about 1000 and 2800 Angstroms.

**VDT**:  Video Display Terminals (see VI-9).

**VDU**:  Video Display Unit (see VI-9).

**Ventilation:**  See VI-10.

**Vibration:**  The alternating motion of a body with respect to a reference point[77] (see VI-13).

**Vibration White Finger Syndrome:**  An occupational disease caused by overexposure of the finger and hands to specific vibration frequencies and intensities.  The disease, characterized by vascular, temperature dependent instability in the fingers, is similar to Raynaud's phenomenon.

$V_R$:  Pulmonary ventilation rate (units vary).

**W:**  SI symbol for the watt.

**watt:**  SI derived unit of power equal to one J/s.

**WBGT:**  Wet Bulb Globe Temperature (see VI-6).

**WBT:**  Wet Bulb Temperature (see VI-6).

**WHO:**  World Health Organization.

**X:**  Reflectance.

**°C:**  Degrees centigrade.

**°K:**  Degrees Kelvin or degrees relative to absolute zero (-273.2°K).

$\lambda$:  Wavelength of electromagnetic radiation.

$\pi$:  The ratio of the circumference to the diameter of any circle (about 3.14159).

# REFERENCES

1. Pub.L No. 101-336, 104 Stat 327 (July 26, 1990).

2. Carl Zenz, M.D., *Occupational Medicine Principles and Applications*, 2nd ed. (Chicago: Year Book Medical Publishers, Inc., 1988), 110.

3. Barbara A. Plog, ed., *Fundamentals of Industrial Hygiene*, 3rd ed. (Chicago: National Safety Council, 1988), 22.

4. Ibid., 613.

5. Levy and Wegman, *Occupational Health, Recognizing and Preventing Work-Related Disease*, 2nd ed. (Boston: Little, Brown and Co., 1988), 31.

6. John M. Burns, M.D., "The Corporate Physician as a Health Management Leader," *J.O.M.* 33, No. 3 (March 1991): 335.

7. Thomas J. McDonagh, M.D., "The Physician as Manager," *J.O.M.* 24, No. 2 (Feb. 1982): 99.

8. Thomas J. McDonagh, M.D., "Management of an Occupational Health Program Within an Industrial Setting," *J.O.M.* 26, No.4 (April 1984): 263.

9. N.S. Stearns, M.D. and E.B. Roberts, "Why Do Occupational Physicians Need to Be Better Managers?," *J.O.M.* 24, No.3 (March 1982): 219.

10. Mary Schmitz and W.C. Courtwright, "Case Study: The Corporate Physician in a Managed Care Environment," *J.O.M.* 33, No. 3 (March 1991): 351.

11. A. Ward Gardner, M.D., *Current Approaches to Occupational Medicine, ref. The Doctor in Industry* (British Medical Association - 1975), (Bristol: John Wright and Sons, Ltd., 1979), 232.

12. Jane A. Lee, R.N., *The New Nurse in Industry* (NIOSH, Pub. No. 78-143, January 1978), 3.

13. William N. Rom, M.D., Ed., "Guide to the Conduct of a Health Hazard Evaluation," in *Environmental and Occupational Medicine*, (Boston: Little, Brown and Co., 1983), 13-19.

14. Barbara A. Plog, ed., *Fundamentals of Industrial Hygiene*, 3rd ed. (Chicago: National Safety Council, 1988), 7-8.

15. William N. Rom, M.D., Ed., "Guide to the Conduct of a Health Hazard Evaluation - Conduct of a Site Visit or Walk-through" in *Environmental and Occupational Medicine*, (Boston: Little, Brown and Co., 1983), 15-19.

16. AIHA, *Standards, Interpretations and Audit Criteria for Performance of Occupational Health Programs*, (Akron: OHSPAC/American Industrial Hygiene Association, 1985).

17. American Medical Association, *Surveying Hazards in the Occupational Environment - The Role of Plant Tours by Physicians in the Development of an Occupational Health and Safety Program*, (Chicago: AMA, May 1976). Reprinted from *Archives of Environmental Health* 10 (Jan. 1965), 115-130.

18. W. A. Burgess, *Recognition of Health Hazards in Industry* (New York: Wiley-Interscience Publications, John Wiley and Sons, 1981).

19. J.M. Harrington, M.D. and F.S. Gill, "Workplace Hazard Control, Sources of Emission" in *Occupational Health*, 2nd ed. (Oxford: Blackwell Scientific Publications, 1987), 228-229.

20. D. H. Collings Jr., M.D., "Examining the 'Occupational' in Occupational Medicine," *J.O.M.* 26, No.7 (July 1984): 510.

21. J. E. Muscat and M.S. Huncharek, "Causation and Disease: Biomedical Science in Toxic Tort Litigation," J.O.M., 31, No. 12, (December 1989): 997.

22. S. Kusnetz and M.R. Hutchinson, *A Guide to the Work-Relatedness of Disease*, (NIOSH, pub. No. 79-216, January 1979).

23. J. P. Kornberg, M.D., "Occupational Medical Causality Assessment," *Provider Pulse, Colorado Compensation Insurance Authority*, (Winter 1991).

24. J. P. Kornberg, M.D., "Occupational Stress - A Specialist's Point of View," (Safety Management Executive Briefing, The Merritt Company, Santa Monica, CA, June 1986).

25. N. M. Hadler, M.D., "Occupational Illness - The Issue of Causality," *J.O.M.* 26, No. 8 (August 1984): 587-593.

26. C. Damme, J.D., "Diagnosing Occupational Disease: A New Standard of Care," *J.O.M.* 20, No.4 (April 1978): 251.

27. R. D. Finucane, M.D. and T.J. McDonagh, M.D., "Medical Information Systems Roundtable," *J.O.M.* 24, No.10 (Oct. 1982): 781-866.

28. Pamela Kerr, "Recording Occupational Health Data for Future Analysis," *J.O.M.* 20, No.3 (March 1978): 197-203.

29. C. P. Wen et. al., "A Result-Oriented Medical Information System," *J.O.M.* 26, No.5 (May 1984): 386-391.

30. R. H. Stoller LTC, "Occupational Medicine: A Proposed Systematic Approach and Model Algorithm for the Physician," *J.O.M.* 25, No.11 (Nov. 1983): 823-828.157

31. G. C. Lowry and R.C. Lowry, *Handbook of Hazard Communication and OSHA Requirements* (Chelsea: Lewis Publishers, 1985) (n.b. Editions of this book are updated with changes in the Federal standard) Ref. U.S. OSHA "Hazard Communication Standard," (29 CFR 1910.1200).

32. ACGIH, *1990-1991 Threshold Limit Values for Chemical Substances and Physical Agents and Biological Exposure Indices.* (Cincinnati, Ohio: American Conference of Governmental Industrial Hygienists, 1990), 4.

33. Barbara A. Plog, ed., *Fundamentals of Industrial Hygiene*, 3rd ed (Chicago: National Safety Council, 1988), 20-21.

34. ACGIH, *1990-1991 Threshold Limit Values for Chemical Substances and Physical Agents and Biological Exposure Indices.* (Cincinnati, Ohio: American Conference of Governmental Industrial Hygienists, 1990), 9.

35. Ibid. 6.

36. Ibid. 7.

37. NIOSH, *NIOSH Pocket Guide to Chemical Hazards*, (Cincinnati: National Institute of Occupational Safety and Health Pub. No. 90-117, June 1990), 4.

38. Department of Labor, OSHA 29 CFR Part 1910; *Air Contaminants*, Final Rule, Federal Register, Vol. 54, No. 12, Thursday, January 19, 1989, Summary, 2332.

39. Ibid. 2384.

40. NIOSH, *NIOSH Pocket Guide to Chemical Hazards*, (Cincinnati: National Institute of Occupational Safety and Health Publication No. 90-117, June 1990), 5.

41. ACGIH, *1990-1991 Threshold Limit Values for Chemical Substances and Physical Agents and Biological Exposure Indices* (Cincinnati: American Conference of Governmental Industrial Hygienists, 1990), 53.

42. Ibid. 59.

43. Ibid. 43.

44. Ibid. 45.

45. William C. Hinds, *Aerosol Technology Properties, Behavior and Measurement of Airborne Particles* (New York: John Wiley and Sons, 1982), 220.

46. M. Lippman, "Recent Advances in Respiratory Tract Particle Deposition" in *Advances in Modern Environmental Toxicology, Vol. III. Occupational and Industrial Hygiene; Concepts and Methods*, Esmen, N.A. and Mehlman, M.A. ed. (Princeton, N.J.: Princeton Scientific Publishers, Inc. 1984), 87.

47. Board on Toxicology & Environmental Health Hazards, *Drinking Water and Health*, Vol. 4, (Washington, D.C.: National Academy Press, 1982), 179.

48. NIOSH, *Work Practice Guide for Manual Lifting*, NIOSH (Cincinnati, Ohio: National Institute of Occupational Safety and Health, Pub. No. 81-122, 1981).

49. Don Chaffin, *Occupational Biomechanics*, 2nd ed. (New York: John Wiley and Sons, Inc., 1991).

50. J. R. Wilson, ed. *Evaluation of Human Work* (London: Taylor and Francis, 1990).

51. S. Konz, *Work Design: Industrial Ergonomics*, 2nd ed. (New York: John Wiley and Sons, 1982).

52. NIOSH, *Criteria for a Recommended Standard Occupational Exposure to Hot Environments*, revised, (Cincinnati, Ohio: NIOSH Pub. No. 86-113, 1986), 101-102, 55.

53. Ibid. 105.

54. Barbara A. Plog, ed., *Fundamentals of Industrial Hygiene*, 3rd ed (Chicago: National Safety Council, 1988), 270.

55. A.M. Best Company, *Best's Safety Directory*, (Oldwich, New Jersey: A.M. Best Company, 1991), 1677.

56. Ibid. 1679-1681.

57. Ibid. 524.

58. A. D. Schwope et al., *Guidelines for the Selection of Chemical Protective Clothing.* (Cincinnati, Ohio: American Conference of Governmental Industrial Hygienists, Feb. 1987).

59. K. Forsberg and S.Z. Mansdorf, *Quick Selection Guide to Chemical Protective Clothing*, 3rd ed. (New York: Van Nostrand Reinhold, 1989).

60. A.M. Best Company, *Best's Safety Directory*. (Oldwich, New Jersey: A.M. Best Company, 1991), 484.

61. NIOSH, *A Report on the Performance of Men's Safety-Toe Footwear* (Cincinnati, Ohio: NIOSH Pub. No. 77-113, July 1976).

62. NIOSH, *Women's Safety-Toe Footwear* (Cincinnati, Ohio: NIOSH Pub. No. 76-199., July 1976).

63. A.M. Best Company, *Best's Safety Directory*. (Oldwich, New Jersey: A.M. Best Company, 1991), 367.

64. Ibid., 1177.

65. AMA, *Guides to the Evaluation of Permanent Impairment*, 3rd ed. Revised, (Chicago: American Medical Association, 1990), 174.

66. A.M. Best Company, *Best's Safety Directory*. (Oldwich, New Jersey: A.M. Best Company, 1991), 1181.

67. B. Ballantyne and P.H. Schwabe, *Respiratory Protection, Principles and Applications*, (London:  Year Book Medical Publishers, Inc. in arrangement with Chapman and Hall, Ltd., 1981), 30.

68. Occupational Safety and Health Act, Title 29, Chapter XVII, Occupational Safety and Health Administration Subpart I - Personal Protection Equipment, Section 1910.134(a)(10), August 27, 1971.

69. ACGIH, *1990-1991 Threshold Limit Values for Chemical Substances and Physical Agents and Biological Exposure Indices*, (Cincinnati, Ohio: American Conference of Governmental Industrial Hygienists, 1990), 115-117.

70. P. A. Howarth, "Assessment of the Visual Environment," in *Evaluation of Human Work*, Wilson, J.R. ed. (London:  Taylor and Francis, 1990), 366.

71. M. David Egan, *Concepts in Architectural Lighting* (New York: McGraw Hill Publishing Co., 1983), 227.

72. P. A. Howarth, "Assessment of the Visual Environment," in *Evaluation of Human Work*, Wilson, J.R. ed. (London:  Taylor and Francis, 1990), 367.

73. Ibid. 365.

74. Barbara A. Plog, ed., *Fundamentals of Industrial Hygiene*, 3rd ed. (Chicago:  National Safety Council, 1988), 238.

75. P.A. Howarth, "Assessment of the Visual Environment," in *Evaluation of Human Work*, Wilson, J.R. ed. (London:  Taylor and Francis, 1990) 369.

76. S. Konz, *Work Design:  Industrial Ergonomics*, 2nd ed. (New York: John Wiley and Sons, 1982), 357.

77. S.H. Rodgers, ed. (Human Factors Section of Eastman Kodak Co.), *Ergonomic Design for People at Work*, Vol. I (New York: Van Nostrand Reinhold Co., 1983), 233.

78. F. Dy, *Visual Display Units: Job Content and Stress in Office Work* (Geneva: International Labour Office, 1985), 42-43.

79. WHO, *Visual Display Terminals and Worker's Health* (Geneva: World Health Organization, Offset Publication No. 99, 1987), 13-14.

80. AIHA, Ergonomics and Non-Ionizing Radiation Committees, *Health and Ergonomic Considerations of Video Display Units*, (Symposium Proceedings, Sheraton Denver Tech Center, March 1-2, 1982), 40.

81. Barbara A. Plog, ed., *Fundamentals of Industrial Hygiene*, 3rd ed. (Chicago: National Safety Council, 1988), 231.

82. WHO, *Lasers and Optical Radiation, Environmental Health Criteria 23* (Geneva: World Health Organization, 1982), 25.

83. G. D. Clayton, ed., *Patty's Industrial Hygiene and Toxicology*, 4th ed., Vol. I, Part A, General Principles, (New York: John Wiley and Sons, Inc., 1991), 664.

84. ACGIH, *1990-1991 Threshold Limit Values for Chemical Substances and Physical Agents and Biological Exposure Indices* (Cincinnati, Ohio: American Conference of Governmental Industrial Hygienists, 1990), 114.

85. G. D. Clayton, ed., *Patty's Industrial Hygiene and Toxicology*, 4th ed., Vol. I, Part A, General Principles (New York: John Wiley and Sons, Inc., 1991), 671.

86. Barbara A. Plog, ed., *Fundamentals of Industrial Hygiene*, 3rd ed. (Chicago: National Safety Council, 1988), 240.

87. NIOSH/OSHA. *Radio Frequency (RF) Sealers and Heaters: Potential Health Hazards and Their Prevention*, Current Intelligence Bulletin 33, December 4, 1979, (Cincinnati, Ohio: NIOSH Pub. No. 80-107, 1979), 4.

88. WHO, *Radio Frequency and Microwaves, Environmental Health Criteria 16*, (Geneva: World Health Organization, 1981), 10.

89. Ibid. 17.

90. Ibid. 82.

91. ACGIH, *1990-1991 Threshold Limit Values for Chemical Substances and Physical Agents and Biological Exposure Indices* (Cincinnati, Ohio: American Conference of Governmental Industrial Hygienists, 1990), 106-107.

92. Barbara A. Plog, ed., *Fundamentals of Industrial Hygiene*, 3rd ed. (Chicago: National Safety Council, 1988), 242.

93. NIOSH, *Laser Hazard Classification Guide* (Cincinnati, Ohio: NIOSH Pub. No. 76-183, July 1976), 3.

94. ACGIH, *1990-1991 Threshold Limit Values for Chemical Substances and Physical Agents and Biological Exposure Indices* (Cincinnati, Ohio: American Conference of Governmental Industrial Hygienists, 1990), 100.

95. P. J. Walsh et al. ed. *Indoor Air Quality* (Boca Raton, Florida: CRC Press, 1984).

96. J. M. Samet and J. D. Spengler, ed., *Indoor Air Pollution, A Health Perspective* (Baltimore: The Johns Hopkins University Press, 1991).

97. Thad Godish, *Indoor Air Pollution Control* (Chelsea, Michigan: Lewis Publishers, 1990).

98. R. B. Gammage and S.V. Kaye, *Indoor Air and Human Health* (Chelsea, Michigan: Lewis Publishers, 1985).

99. R. A. Wadden and P.A. Scheff, *Indoor Air Pollution, Characterization, Prediction and Control* (New York: Wiley Interscience, 1983).

100. ACGIH Committee on Industrial Ventilation, *Industrial Ventilation - A Manual of Recommended Practice*, 17th ed. (Lansing, Michigan: American Conference of Governmental Industrial Hygienists, 1982).

101. H. J. McDermott, *Handbook of Ventilation Contaminant Control*. 2nd ed. (Boston: Butterworth Publishers, 1985), 16-17.

102. N. Sax and R. Lewis, *Hawley's Condensed Chemical Dictionary*, 11th ed. (New York: Van Nostrand Reinhold Co., 1987), 1083.

103. G. Weiss, *Hazardous Chemicals Data Book* (Park Ridge, New Jersey: Noyes Data Corporation, 1986), 604.

104. L. F. Yerges, *Sound, Noise and Vibration Control* (New York: Van Nostrand Reinhold Co., 1978), 7.

105. A. Peterson, *Handbook of Noise Measurement*, 9th ed. (Concord, Massachusetts: GENRAD, 1980), 7.

106. J. Sataloff et al., *Hearing Loss*. 2nd ed. (Philadelphia: J.B. Lippincott Co., 1980), 379.

107. A. Peterson, *Handbook of Noise Measurement*, 9th ed. (Concord, Massachusetts: GENRAD, 1980), 8.

108. Ibid. 9.

109. ACGIH, *1990-1991 Threshold Limit Values for Chemical Substances and Physical Agents and Biological Exposure Indices* (Cincinnati, Ohio: American Conference of Governmental Industrial Hygienists, 1990), 104.

110. Donald E. Wasserman, "Human Aspects of Occupational Vibration." in *Advances in Human Factors/Ergonomics*, 8, (Amsterdam:  Elsevier Science Publishers, 1987).

111. A. J. Brammer and W. Taylor, *Vibration Effects on the Hand and Arm in Industry* (New York:  John Wiley & Sons, 1982).

112. WHO, *Noise and Vibration in the Working Environment* (Geneva: World Health Organization, Occupational Safety & Health Series 33, 1976).

113. NIOSH, *Vibration Syndrome, Current Intelligence Bulletin 38.* (Cincinnati, Ohio:  NIOSH Pub. No. 83-110., March 29, 1983).

114. ACGIH, *1990-1991 Threshold Limit Values for Chemical Substances and Physical Agents and Biological Exposure Indices* (Cincinnati, Ohio:  American Conference of Governmental Industrial Hygienists, 1990), 82-86.

115. NIOSH, *Criteria for a Recommended Standard...Working in Confined Spaces* (Cincinnati, Ohio:  National Institute of Occupational Safety and Health, Pub. No. 80-106, Dec. 1979).

116. NIOSH, *NIOSH Alert: Request for Assistance in Preventing Occupational Fatalities in Confined Spaces* (Cincinnati, Ohio: National Institute of Occupational Safety and Health, Pub. No. 86-110, January 1986).

117. NIOSH: *A Guide to Safety in Confined Spaces* (Cincinnati, Ohio: National Institute of Occupational Safety and Health, Pub. No. 87-113, July 1987).

118. NIOSH, *NIOSH Alert: Request for Assistance in Preventing Occupational Fatalities in Confined Spaces* (Cincinnati, Ohio: National Institute of Occupational Safety and Health, Pub. No. 86-110, January 1986), 2.

119. NIOSH, *Criteria for a Recommended Standard...Working in Confined Spaces* (Cincinnati, Ohio: National Institute of Occupational Safety and Health, Pub. No. 80-106, Dec. 1979), 1.

120. R. C. Weast. ed, *CRC Handbook of Chemistry and Physics*, 67th ed. (Boca Raton, Florida: CRC Press Inc., 1986-87), F-289.

121. Ibid. F-68.

122. Ibid. F-286.

123. Ibid, F-234.

124. Ibid, F-240.

125. Ibid, F-283.

# INDEX

Milton Keynes UK
Ingram Content Group UK Ltd.
UKHW051952071024
449327UK00026B/2283